THE NATURE OF SCIENCE

THE NATURE OF SCIENCE

Classical and Contemporay Readings

Edited by
Byron E. Wall
York University

Wall & Emerson, Inc.
Toronto, Ontario • Dayton, Ohio

Orders for this book may be sent in any of the following ways:

By mail to:

Wall & Emerson, Inc.
Six O'Connor Drive
Toronto, Ontario, Canada
M4K 2K1

By telephone, facsimile, email, or via the world wide web:

Telephone: (416) 467-8685
Fax: (416) 696-2460
E-mail: wall@wallbooks.com
Website: www.wallbooks.com

This work is Unit 1 of Science in Society: Classical and Contemporary Readings, edited by Byron E. Wall (Toronto: Wall & Emerson, Inc. 1989).

Cover:

Scientific equipment in the Swiss Polytechnic in Zurich where Einstein studied from 1896 to 1899.

Printed in Canada.

ISBN 0-921332-36-X

4 5 6 7 8 9 03 02 01 00 99 98

Table of Contents

Preface

This volume is Unit 1 of the anthology *Science in Society: Classical and Contemporary Readings*, edited by Byron E. Wall (Toronto: Wall & Emerson, Inc., 1989). That work is intended as a reader for courses that consider science and technology in both an historical and a contemporary social context. This excerpted volume addresses the historical and philosophical issues, specifically, of science itself, and is intended for introductory courses that consider the nature and development of science.

The selections are a combination of classical documents or commentaries from the history of science and recent pieces, chosen for their readability and interest. Articles marked in the table of contents and on their first pages with an asterisk (*) are more challenging and are intended for the better prepared students. Alternatively, these articles might be assigned as special projects or as starting points for essays.

Each article is followed by questions and topics for discussion and writing. There is considerable variation in the depth and complexity of the questions, generally with the easier questions first and the more demanding ones further on down the list. Many of these questions and topics would make suitable essay or research paper topics.

This volume is probably best used as a stimulus for class discussion or for background reading. Instructors should be aware that the articles have been selected in order that they stand on their own and are interesting by themselves, rather than that they "cover" the topic. Therefore many issues that might have been included are skipped entirely, or mentioned only in passing. For example, there is very little about the development of chemistry, and nothing about DNA and other aspects of biology in the twentieth century.

This book can be used as a supplementary reader along with a survey text that covers the major topics in the history of science. A good choice would be *Children of Prometheus: A History of Science and Technology* by James MacLachlan (Toronto: Wall & Emerson, Inc., 1989).

At the first Solvay Congress in Brussels in 1911. Most of Europe's top physicists of the early twentieth century are in the picture, including Marie Curie, seated, and Albert Einstein, standing, second from right.

Issue

The Nature of Scientific Knowledge

It would seem to be a good idea to begin a book that looks at the context of science with a definition of science itself. But defining something as complex and far-reaching as science is no easy task. A whole branch of philosophy—the philosophy of science—is devoted to analysing the nature of scientific knowledge and its reasoning processes. The literature of the philosophy of science is vast, difficult, and full of areas of disagreement. Yet ordinary people speak of science frequently and at least *think* they are making themselves understood.

The articles in this section are intended to help clarify some of the major issues in defining the nature of science without getting too far afield in some of the more subtle points. William Hanna's article, "The Nature of Scientific Reasoning," lays out some of the main issues that are discussed in greater detail in other articles in this unit and which are referred to throughout the book. Isaac Asimov's "The Eureka Phenomenon" examines the creative process in science, particularly as it applies to finding new relationships and hypotheses. Berton Roueché illustrates deductive reasoning in "The Orange Man," where established scientific knowledge is used to diagnose an illness. Carl Sagan writes about science as a special kind of thinking about the world and considers the limits of human knowledge in "Can We Know the Universe?" Thomas Henry Huxley's lecture, "We All Use the Scientific Method," shows how the processes of observation, induction, hypothesis formation and evaluation that are considered hallmarks of the scientific method are also everyday tools of common sense thinking. An example of the careful analysis of scientific thinking by philosophers of science is Carl G. Hempel's "Scientific Inquiry: Invention and Test." Hempel shows how carefully chosen discriminating observations and experiments can enable a scientist to posit a likely hypothesis and then test it by testing its logical consequences. Finally, to illustrate the real scientist at work, Marie Curie's lecture, "The Discovery of Radium," describes her work investigating the nature of radioactivity and the methods she employed. However, more than the methods, what she really conveys is the joy of scientific work and the beauty of scientific knowledge.

The Nature of Scientific Reasoning

William R. Hanna

Just what *is* science anyway? If we want to talk sensibly about science in society, we had better have some clear idea what we mean by "science" and what makes a subject or statement "scientific." This article, by William R. Hanna of Humber College in Toronto, explores the basic issues.

What is known as modern science came into existence only a few hundred years ago, yet it has had a profound impact on almost every facet of our lives. The western world is inundated with the practical applications of scientific knowledge. These applications have led to improvements in farming, manufacturing, transportation, health—the list goes on and on. But the technical advances based on scientific knowledge have not only led to an easier and more comfortable life. There is also a negative side represented by such things as: the possibility of nuclear destruction, global pollution, changes to the climate, etc. In addition to improvements and threats, there are whole new fields being developed that open new possibilities and present unknown consequences. Computer technology, genetic engineering and molecular biology are all new fields that have only recently emerged. The impact of science and technology on our daily existence is of such magnitude that it would be folly to remain ignorant of both new developments and of the nature of scientific reasoning and the scientific process.

Given the way in which these advances, problems and benefits impinge upon our consciousness through the products of applied science, it is easy to confuse technological achievements with "science" itself. When not bewildered by the confusing array of technological "toys," public perception tends to focus on the complexity of scientific procedures and on the gulf separating the "real" world from the world in which scientists work. We tend to forget that science is a very human activity—a way of thinking. Over the last three hundred years, however, this way of thinking has dramatically expanded the range of human knowledge. It has both created new problems and offered tantalizing new visions. To understand the problems and to share in the visions, it is necessary to look more closely at the nature and limits of scientific reasoning.

* * *

When we talk about science, we normally refer to the activity of investigating nature. This activity must be faithful to the evidence gathered through careful observation and experimentation and it must follow certain rules of logic. Science studies particular aspects of our knowledge upon which, in principle, there can be universal agreement. It is perceived as seeking, and getting, maximum consensus.

The word itself refers to both a body of knowledge and the set of rules by which this knowledge is gathered. A dictionary definition of science might state that science is: "Knowledge of the world ascertained by observation, critically examined and classified systematically under general principles." This knowledge is also presumed to explain or encompass past discoveries and to predict future developments or occurrences. It is measurable insofar as the relation between hypotheses and their implications may be evaluated. A scientist, therefore, generates ideas which can then be tested through observation. The emphasis is not on the generation of ideas, but on their verification through testing.

Although there is room in science for luck and

guesses and "bolts from the blue," science is usually perceived as disciplined investigation. It is a series of questions followed by an attempt to answer these questions through systematic steps. Facts are observed, hypotheses are formed, and hypotheses are tested through controlled experimentation. Verified results of such methodology become the generalizations known as scientific theories or laws.

While the practical applications of science have certainly stimulated its development, curiosity plays an important role as well. There is a quest for general truths—truths that are illustrated by particular facts. These facts may be known by direct observation. A ball falls when released. The moon changes position in the sky. These are matters open to inspection. Science is not a simple record of such phenomena: it is an attempt to understand them. It is an effort to formulate general laws that describe underlying patterns and their interrelation. When a scientist observes an event, he looks for the natural law which governs that event and for the underlying principle that explains how it fits into the greater order of things. The role of theories is to explain facts. Galileo (1564–1642) formulated the laws of terrestrial mechanics to explain the motion of bodies on or near the earth. Kepler (1571–1630) formulated the laws of celestial mechanics to explain planetary motion. It was up to Newton (1642–1727) to unify and explain them through his Theory of Universal Gravitation and the Laws of Motion.

Strictly speaking, then, scientific method connects present observation to past experience. This allows for the generation of a hypothesis from which predictions can be made. Findings or evidence are then subject to the verdict of experience. The merit of the hypothesis must be assessed. The basic idea is straightforward. Starting with some knowledge, we infer that, if X then Y. Then, given some figures, we predict the likely result as exactly as possible. The prediction is put to the test of experience and is either upheld or refuted. If refuted, we either discard the hypothesis or try an amended version.

As mentioned above, hypotheses must be relevant to the facts they are intended to explain. Conversely, the fact in question must be deducible from the proposed hypothesis. Hypotheses must also be testable—there must be the possibility of making observations that either confirm or disprove the hypothesis. They must also be compatible and consistent with other, "established" hypotheses. Sometimes, there may be two or more hypotheses which are relevant, compatible and subject to verification. Scientists then look to predictive power and simplicity to choose among conflicting theories. The greater the range of observable facts which can be deduced from a hypothesis and the simpler the explanation, the more likely it is that a given hypothesis will be accepted.

* * *

It is important to remember that scientific laws, theories and hypotheses are not certainties. Invariability in science is only invariable until proven otherwise. Laws of substance, for example, describe such things as elements and compounds. At one time, the freezing and boiling temperatures of water were thought to be invariable. It was discovered, however, that the application of pressure may effect a significant change on these. The law that explains these properties of water had to be modified. Scientists can never be certain that important factors may not have been considered. Generally speaking, laws state invariable relations and regularities about nature which are subject to empirical confirmation. Theories attempt to explain laws. The history of science is littered with the corpses of theories that could not explain newly discovered laws. Theories are necessary but fallible inventions of the mind, constructed in the effort to make sense of the observable world. They are a valuable tool of inquiry, but they are never certainties.

We have made reference to both deduction and induction. These are both elements of logic, and logic plays a key role in scientific method. In essence, logic is the study of the principles of correct reasoning. It analyses arguments and the clarity of their expression. Logic does not determine the quality of evidence (facts) nor the quality of conclusions. Its role is to determine the quality of the relation between evidence and conclusion—arguments. Scientific theories are built on arguments, and the success of a theory is based, in part, upon the soundness of its logic.

Science is often considered to be very methodical and pedestrian: observation of data, determination of relations and regularities, divination of organizing and underlying patterns. This is a misrepresentation of scientific method. While the method is indeed rigorous, and the testing is perhaps routine, the very fact that hypotheses and theories are not certainties indicates their conjectural nature. They are educated guesses. There is conjecture involved in every use of evidence. Underlying patterns are not written on every event or experience. The act of discovery is the art of informed speculation; the proposing of answers to questions beginning "What if..."

Speculation is one of the distinguishing features of humankind. In this sense, science is a tool used more or less rigorously by everyone. It has limits and limitations. Its products are both a curse and a boon. Offering both a pathway to knowledge and the challenge of discovery, science is not likely to be discarded because of the consequences of its applications. The technological manifestations of science are its products. If products malfunction, create hazards, etc., you don't blame the tools used in production—you look to the manufacturer.

Questions and Topics for Discussion and Writing

1. Give a definition of science and a definition of another kind of knowledge which you would not classify as science.
2. Make lists of subject statements, and theories in three categories: (1) those which are clearly scientific, (2) those which are clearly not scientific, and (3) those which could be either scientific or not.
3. Discuss some of the items selected in category (3) in question 2 above. Explain how each item could be called scientific and how it could be called non-scientific.

William R. Hanna

The Eureka Phenomenon

Isaac Asimov

Discussions of the nature of science usually concentrate on aspects of science which *differ* from non-science and which have to do with verification and reasoning procedures. But while science does have special means by which facts become established and accepted, there would be no body of scientific knowledge at all without the creative process that conceived the scientific propositions to begin with. In this article prolific science writer Isaac Asimov discusses the creative process in science as revealed in some famous examples.

In the old days, when I was writing a great deal of fiction, there would come, once in a while, moments when I was stymied. Suddenly, I would find I had written myself into a hole and could see no way out. To take care of that, I developed a technique which invariably worked.

It was simply this—I went to the movies. Not just any movie. I had to pick a movie which was loaded with action but which made no demands on the intellect. As I watched, I did my best to avoid any conscious thinking concerning my problem, and when I came out of the movie I knew exactly what I would have to do to put the story back on the track.

It never failed.

In fact, when I was working on my doctoral dissertation, too many years ago, I suddenly came across a flaw in my logic that I had not noticed before and that knocked out everything I had done. In utter panic, I made my way to a Bob Hope movie—and came out with the necessary changes in point of view.

It is my belief, you see, that thinking is a double phenomenon like breathing.

You can control breathing by deliberate voluntary action: you can breathe deeply and quickly, or you can hold your breath altogether, regardless of the body's needs at the time. This, however, doesn't work well for very long. Your chest muscles grow tired, your body clamors for more oxygen, or less, and you relax. The automatic involuntary control of breathing takes over, adjusts it to the body's needs and unless you have some respiratory disorder, you can forget about the whole thing.

Well, you can think by deliberate voluntary action, too, and I don't think it is much more efficient on the whole than voluntary breath control is. You can deliberately force your mind through channels of deductions and associations in search of a solution to some problem and before long you have dug mental furrows for yourself and find yourself circling round and round the same limited pathways. If those pathways yield no solution, no amount of further conscious thought will help.

On the other hand, if you let go, then the thinking process comes under automatic involuntary control and is more apt to take new pathways and make erratic associations you would not think of consciously. The solution will then come while you *think* you are *not* thinking.

The trouble is, though, that conscious thought involves no muscular action and so there is no sensation of physical weariness that would force you to quit. What's more, the panic of necessity tends to force you to go on uselessly, with each added bit of useless effort adding to the panic in a vicious cycle.

It is my feeling that it helps to relax, deliberately, by subjecting your mind to material complicated enough to occupy the voluntary faculty of thought, but superficial enough not to engage the deeper involuntary one. In my case, it is an action movie; in your case, it might be something else.

I suspect it is the involuntary faculty of thought that gives rise to what we call "a flash of intuition," something that I imagine must be merely the result of unnoticed thinking.

Perhaps the most famous flash of intuition in the

history of science took place in the city of Syracuse in third-century B.C. Sicily. Bear with me and I will tell you the story.

About 250 B.C., the city of Syracuse was experiencing a kind of Golden Age. It was under the protection of the rising power of Rome, but it retained a king of its own and considerable self-government; it was prosperous; and it had a flourishing intellectual life.

The king was Hieron II, and he had commissioned a new golden crown from a goldsmith, to whom he had given an ingot of gold as raw material. Hieron, being a practical man, had carefully weighed the ingot and then weighed the crown he received back. The two weights were precisely equal. Good deal!

But then he sat and thought for a while. Suppose the goldsmith had subtracted a little bit of the gold, not too much, and had substituted an equal weight of the considerably less valuable copper. The resulting alloy would still have the appearance of pure gold, but the goldsmith would be plus a quantity of gold over and above his fee. He would be buying gold with copper, so to speak, and Hieron would be neatly cheated.

Hieron didn't like the thought of being cheated any more than you or I would, but he didn't know how to find out for sure if he had been. He could scarcely punish the goldsmith on mere suspicion. What to do?

Fortunately, Hieron had an advantage few rulers in the history of the world could boast. He had a relative of considerable talent. The relative was named Archimedes and he probably had the greatest intellect the world was to see prior to the birth of Newton.

Archimedes was called in and was posed the problem. He had to determine whether the crown Hieron showed him was pure gold, or was gold to which a small but significant quantity of copper had been added.

If we were to reconstruct Archimedes' reasoning, it might go as follows. Gold was the densest known substance (at that time). Its density in modern terms is 19.3 grams per cubic centimeter. This means that a given weight of gold takes up less volume than the same weight of anything else! In fact, a given weight of pure gold takes up less volume than the same weight of *any* kind of impure gold.

The density of copper is 8.92 grams per cubic centimeter, just about half that of gold. If we consider 100 grams of pure gold, for instance, it is easy to calculate it to have a volume of 5.18 cubic centimeters. But suppose that 100 grams of what looked like pure gold was really only 90 grams of gold and 10 grams of copper. The 90 grams of gold would have a volume of 4.66 cubic centimeters, while the 10 grams of copper would have a volume of 1.12 cubic centimeters; for a total value of 5.78 cubic centimeters.

The difference between 5.18 cubic centimeters and 5.78 cubic centimeters is quite a noticeable one, and would instantly tell if the crown were of pure gold, or if it contained 10 per cent copper (with the missing 10 per cent of gold tucked neatly in the goldsmith's strongbox).

All one had to do, then, was measure the volume of the crown and compare it with the volume of the same weight of pure gold.

The mathematics of the time made it easy to measure the volume of many simple shapes: a cube, a sphere, a cone, a cylinder, any flattened object of simple regular shape and known thickness, and so on.

We can imagine Archimedes saying, "All that is necessary, sire, is to pound that crown flat, shape it into a square of uniform thickness, and then I can have the answer for you in a moment."

Whereupon Hieron must certainly have snatched the crown away and said, "No such thing. I can do that much without you; I've studied the principles of mathematics, too. This crown is a highly satisfactory work of art and I won't have it damaged. Just calculate its volume without in any way altering it."

But Greek mathematics had no way of determining the volume of anything with a shape as irregular as the crown, since integral calculus had not yet been invented (and wouldn't be for two thousand years, almost). Archimedes would have had to say, "There is no known way, sire, to carry through a non-destructive determination of volume."

"Then think of one," said Hieron testily.

And Archimedes must have set about thinking of one, and gotten nowhere. Nobody knows how long he thought, or how hard, or what hypotheses he considered and discarded, or any of the details.

What we do know is that, worn out with thinking,

Archimedes decided to visit the public baths and relax. I think we are quite safe in saying that Archimedes had no intention of taking his problem to the baths with him. It would be ridiculous to imagine he would, for the public baths of a Greek metropolis weren't intended for that sort of thing.

The Greek baths were a place for relaxation. Half the social aristocracy of the town would be there and there was a great deal more to do than wash. One steamed one's self, got a massage, exercised, and engaged in general socializing. We can be sure that Archimedes intended to forget the stupid crown for a while.

One can envisage him engaging in light talk, discussing the latest news from Alexandria and Carthage, the latest scandals in town, the latest funny jokes at the expense of the country-squire Romans—and then he lowered himself into a nice hot bath which some bumbling attendant had filled too full.

The water in the bath slopped over as Archimedes got in. Did Archimedes notice that at once, or did he sigh, sink back, and paddle his feet awhile before noting the water-slop? I guess the latter. But, whether soon or late, he noticed, and that one fact, added to all the chains of reasoning his brain had been working on during the period of relaxation when it was unhampered by the comparative stupidities (even in Archimedes) of voluntary thought, gave Archimedes his answer in one blinding flash of insight.

Jumping out of the bath, he proceeded to run home at top speed through the streets of Syracuse. He did *not* bother to put on his clothes. The thought of Archimedes running naked through Syracuse has titillated dozens of generations of youngsters who have heard this story, but I must explain that the ancient Greeks were quite lighthearted in their attitude toward nudity. They thought no more of seeing a naked man on the streets of Syracuse, than we would on the Broadway stage.

And as he ran, Archimedes shouted over and over, "I've got it! I've got it! " Of course, knowing no English, he was compelled to shout it in Greek, so it came out, "Eureka! Eureka! "

Archimedes' solution was so simple that anyone could understand it—once Archimedes explained it.

If an object that is not affected by water in any way, is immersed in water, it is bound to displace an amount of water equal to its own volume, since two objects cannot occupy the same space at the same time.

Suppose, then, you had a vessel large enough to hold the crown and suppose it had a small overflow spout set into the middle of its side. And suppose further that the vessel was filled with water exactly to the spout, so that if the water level were raised a bit higher, however slightly, some would overflow.

Next, suppose that you carefully lower the crown into the water. The water level would rise by an amount equal to the volume of the crown, and that volume of water would pour out the overflow and be caught in a small vessel. Next, a lump of gold, known to be pure and exactly equal in weight to the crown, is also immersed in the water and again the level rises and the overflow is caught in a second vessel.

If the crown were pure gold, the overflow would be exactly the same in each case, and the volume of water caught in the two small vessels would be equal. If, however, the crown were of alloy, it would produce a larger overflow than the pure gold would and this would be easily noticeable.

What's more, the crown would in no way be harmed, defaced, or even as much as scratched. More important, Archimedes had discovered the "principle of buoyancy."

And was the crown pure gold? I've heard that it turned out to be alloy and that the goldsmith was executed, but I wouldn't swear to it.

How often does this "Eureka phenomenon" happen? How often is there this flash of deep insight during a moment of relaxation, this triumphant cry of "I've got it! I've got it! " which must surely be a moment of the purest ecstasy this sorry world can afford?

I wish there were some way we could tell. I suspect that in the history of science it happens *often*; I suspect that very few significant discoveries are made by the pure technique of voluntary thought; I suspect that voluntary thought may possibly prepare the ground (if even that), but that the final touch, the real inspiration, comes when thinking is under involuntary control.

But the world is in a conspiracy to hide the fact. Scientists are wedded to reason, to the meticulous working out of consequences from assumptions to the careful organization of experiments designed to check those consequences. If a certain line of experiments ends nowhere, it is omitted from the final report. If an inspired guess turns out to be correct, it is *not* reported as an inspired guess. Instead, a solid line of voluntary thought is invented after the fact to lead up to the thought, and that is what is inserted in the final report.

The result is that anyone reading scientific papers would swear that *nothing* took place but voluntary thought maintaining a steady clumping stride from origin to destination, and that just can't be true.

It's such a shame. Not only does it deprive science of much of its glamour (how much of the dramatic story in Watson's *Double Helix* do you suppose got into the final reports announcing the great discovery of the structure of DNA? [1]), but it hands over the important process of "insight," "inspiration," "revelation" to the mystic.

The scientist actually becomes ashamed of having what we might call a revelation, as though to have one is to betray reason—when actually what we call revelation in a man who has devoted his life to reasoned thought, is after all merely reasoned thought that is not under voluntary control.

Only once in a while in modern times do we ever get a glimpse into the workings of involuntary reasoning, and when we do, it is always fascinating. Consider, for instance, the case of Friedrich August Kekule von Stradonitz.

In Kekule's time, a century and a quarter ago, a subject of great interest to chemists was the structure of organic molecules (those associated with living tissue). Inorganic molecules were generally simple in the sense that they were made up of few atoms. Water molecules, for instance, are made up of two atoms of hydrogen and one of oxygen (H_2O). Molecules of ordinary salt are made up of one atom of sodium and one of chlorine (NaCl), and so on.

Organic molecules, on the other hand, often contained a large number of atoms. Ethyl alcohol molecules have two carbon atoms, six hydrogen atoms, and an oxygen atom (C_2H_6O); the molecule of ordinary cane sugar is $C_{12}H_{22}O_{11}$, and other molecules are even more complex.

Then, too, it is sufficient, in the case of inorganic molecules generally, merely to know the kinds and numbers of atoms in the molecule; in organic molecules, more is necessary. Thus, dimethyl ether has the formula C_2H_6O, just as ethyl alcohol does, and yet the two are quite different in properties. Apparently, the atoms are arranged differently within the molecules—but how to determine the arrangements?

In 1852, an English chemist, Edward Frankland, had noticed that the atoms of a particular element tended to combine with a fixed number of other atoms. This combining number was called "valence." Kekule in 1858 reduced this notion to a system. The carbon atom, he decided (on the basis of plenty of chemical evidence) had a valence of four; the hydrogen atom, a valence of one; and the oxygen atom, a valence of two (and so on).

Why not represent the atoms as their symbols plus a number of attached dashes, that number being equal to the valence? Such atoms could then be put together as though they were so many Tinker Toy units and "structural formulas" could be built up.

It was possible to reason out that the structural formula of ethyl alcohol was

```
    H  H
    |  |
 H—C—C—O—H
    |  |
    H  H
```

while that of dimethyl ether was

```
    H     H
    |     |
 H—C—O—C—H
    |     |
    H     H
```

[1] I'll tell you, in case you're curious. None! [Asimov's note.]

In each case, there were two carbon atoms, each with four dashes attached; six hydrogen atoms, each with one dash attached; and an oxygen atom with two ashes attached. The molecules were built up of the same components, but in different arrangements.

Kekule's theory worked beautifully. It has been immensely deepened and elaborated since his day, but you can still find structures very much like Kekule's Tinker Toy formulas in any modern chemical textbook. They represent oversimplifications of the true situation, but they remain extremely useful in practice even so.

The Kekule structures were applied to many organic molecules in the years after 1858 and the similarities and contrasts in the structures neatly matched similarities and contrasts in properties. The key to the rationalization of organic chemistry had, it seemed, been found.

Yet there was one disturbing fact. The well-known chemical benzene wouldn't fit. It was known to have a molecule made up of equal numbers of carbon and hydrogen atoms. Its molecular weight was known to be 78 and a single carbon-hydrogen combination had a weight of 13. Therefore, the benzene molecule had to contain six carbon-hydrogen combinations and its formula had to be C_6H_6.

But that meant trouble. By the Kekule formulas, the hydrocarbons (molecules made up of carbon and hydrogen atoms only) could easily be envisioned as chains of carbon atoms with hydrogen atoms attached. If all the valences of the carbon atoms were filled with hydrogen atoms, as in "hexane," whose molecule looks like this—

```
    H  H  H  H  H  H
    |  |  |  |  |  |
 H—C—C—C—C—C—C—H
    |  |  |  |  |  |
    H  H  H  H  H  H
```

the compound is said to be saturated. Such saturated hydrocarbons were found to have very little tendency to react with other substances.

If some of the valences were not filled, unused bonds were added to those connecting the carbon atoms. Double bonds were formed as in "hexene"—

```
    H  H  H  H  H  H
    |  |  |  |  |  |
 H—C—C—C=C—C—C—H
    |  |        |  |
    H  H        H  H
```

Hexene is unsaturated, for that double bond has a tendency to open up and add other atoms. Hexene is chemically active.

When six carbons are present in a molecule, it takes fourteen hydrogen atoms to occupy all the valence bonds and make it inert—as in hexane. In hexene, on the other hand, there are only twelve hydrogens. If there were still fewer hydrogen atoms, there would be more than one double bond; there might even be triple bonds, and the compound would be still more active than hexene.

Yet benzene, which is C_6H_6 and has eight fewer hydrogen atoms than hexane, is *less* active than hexene, which has only two fewer hydrogen atoms than hexane. In fact, benzene is even less active than hexane itself. The six hydrogen atoms in the benzene molecule seem to satisfy the six carbon atoms to a greater extent than do the fourteen hydrogen atoms in hexane.

For heaven's sake, why?

This might seem unimportant. The Kekule formulas were so beautifully suitable in the case of so many compounds that one might simply dismiss benzene as an exception to the general rule.

Science, however, is not English grammar. You can't just categorize something as an exception. If the exception doesn't fit into the general system, then the general system must be wrong.

Or, take the more positive approach. An exception can often be made to fit into a general system, provided the general system is broadened. Such broadening generally represents a great advance and for this reason, exceptions ought to be paid great attention.

For some seven years, Kekule faced the problem of benzene and tried to puzzle out how a chain of six carbon atoms could be completely satisfied with as few as six hydrogen atoms in benzene and yet be left unsatisfied with twelve hydrogen atoms in hexane.

Nothing came to him!

And then one day in 1865 (he tells the story himself) he was in Ghent, Belgium, and in order to get to some destination, he boarded a public bus. He was tired and, undoubtedly, the droning beat of the horses' hooves on the cobblestones lulled him. He fell into a comatose half-sleep.

In that sleep, he seemed to see a vision of atoms attaching themselves to each other in chains that moved about. (Why not? It was the sort of thing that constantly occupied his waking thoughts.) But then one chain twisted in such a way that head and tail joined, forming a ring—and Kekule woke with a start.

To himself, he must surely have shouted "Eureka," for indeed he had it. The six carbon atoms of benzene formed a ring and not a chain, so that the structural formula looked like this:

```
          H
          |
 H       C       H
   \   /   \\   /
     C       C
     ||      |
     C       C
   /   \   //   \
 H       C       H
         |
         H
```

To be sure, there were still three double bonds, so you might think the molecule had to be very active—but now there was a difference. Atoms in a ring might be expected to have different properties from those in a chain and double bonds in one case might not have the properties of those in the other. At least, chemists could work on that assumption and see if it involved them in contradictions.

It didn't. The assumption worked excellently well. It turned out that organic molecules could be divided into two groups: aromatic and aliphatic. The former had the benzene ring (or certain other similar rings) as part of the structure and the latter did not. Allowing for different properties within each group, the Kekule structures worked very well.

For nearly seventy years, Kekule's vision held good in the hard field of actual chemical techniques, guiding the chemist through the jungle of reactions that led to the synthesis of more and more molecules. Then, in 1932, Linus Pauling applied quantum mechanics to chemical structure with sufficient subtlety to explain just why the benzene ring was so special and what had proven correct in practice proved correct in theory as well.

* * *

Other cases? Certainly.

In 1764, the Scottish engineer James Watt was working as an instrument maker for the University of Glasgow. The university gave him a model of a Newcomen steam engine, which didn't work well, and asked him to fix it. Watt fixed it without trouble, but even when it worked perfectly, it didn't work well. It was far too inefficient and consumed incredible quantities of fuel. Was there a way to improve that?

Thought didn't help, but a peaceful, relaxed walk on a Sunday afternoon did. Watt returned with the key notion in mind of using two separate chambers, one for steam only and one for cold water only, so that the same chamber did not have to be constantly cooled and reheated to the infinite waste of fuel.

The Irish mathematician William Rowan Hamilton worked up a theory of "quaternions" in 1843 but couldn't complete that theory until he grasped the fact that there were conditions under which $p \times q$ was *not* equal to $q \times p$. The necessary thought came to him in a flash one time when he was walking to town with his wife.

The German physiologist Otto Loewi was working on the mechanism of nerve action, in particular, on the chemicals produced by nerve endings. He awoke at 3 A.M. one night in 1921 with a perfectly clear notion of the type of experiment he would have to run to settle a key point that was puzzling him. He wrote it down and went back to sleep. When he woke in the morning, he found he couldn't remember what his inspiration had been. He remembered he had written it down, but he couldn't read his writing.

The next night, he woke again at 3 A.M. with the clear thought once more in mind. This time, he didn't fool around. He got up, dressed himself, went straight to the laboratory and began work. By 5 A.M. he had proved his point and the consequences of his findings became important enough in later years so that in 1936 he received a share in the Nobel prize in medicine and physiology.

Isaac Asimov

How very often this sort of thing must happen, and what a shame that scientists are so devoted to their belief in conscious thought that they so consistently obscure the actual methods by which they obtain their results.

From Isaac Asimov, *The Left Hand of the Election (Garden City, N.Y.: Doubleday, 1971).*

Questions and Topics for Discussion and Writing

1. Why does Asimov want the reader to understand the role of the "Eureka Phenomenon" in science? What is the conception of scientific reasoning that he contrasts it to?
2. Why is it important that the scientist *not* be concentrating on the problem to be solved to make the breakthrough insight possible?
3. Asimov contrasts science to English grammar, which can accommodate exceptions to rules. Why can science not have such exceptions?

The Orange Man

Berton Roueché

An important part of scientific thinking is reasoning by deduction. Deduction is a part of every stage of science, but it is particularly characteristic of applications of a body of scientific knowledge to investigations. The practice of medicine provides especially good examples.

How does a doctor diagnose an illness? What are the means by which the patient's symptoms guide the doctor to the identification of the disease at hand and to the proper treatment? This article by science and medical writer Berton Roueché describes the diagnostic process for a rather bizarre case.

Around eleven-thirty on the morning of December 15, 1960, Dr. Richard L. Wooten, an internist and an assistant professor of internal medicine at the University of Tennessee College of Medicine, in Memphis, was informed by the receptionist in the office he shares with several associates that a patient named (I'll say) Elmo Turner was waiting to see him. Dr. Wooten remembered Turner, but not much about him. He asked the receptionist to fetch him Turner's folder, and then, when she had done so, to send Turner right on in. The folder refreshed his memory. Turner was fifty-three years old, married, and a plumber by trade, and over the past ten years, Dr. Wooten had seen him through an attack of pneumonia and referred him along for treatment of a variety of troubles, including a fractured wrist and hip-joint condition. There were footsteps in the hall. Dr. Wooten closed the folder. The door opened, and Turner—a short, thick, muscular man—came in. Dr. Wooten had risen to greet him, but for a moment he could only stand and stare. Turner's face was orange—a golden, pumpkin orange. So were both his hands.

Dr. Wooten found his voice. He gave Turner a friendly good morning, asked him to sit down, and remarked that it had been a couple of years since their last meeting. Turner agreed that it had. He had been away. He had been working up in Alaska—in Fairbanks. He and his wife were back in Memphis only on a matter of family business. But, being in town, he though he ought to pay Dr. Wooten a visit. There was something that kind of bothered him. Dr. Wooten listened with half an ear. His mind was searching through the spectrum of pathological skin discolorations. There were many diseases with pigmentary manifestations. There was the paper pallor of pituitary disease. There was the cyanotic blue of congenital heart disease. There was the deep Florida tan of thyroid dysfunction. There was the jaundice yellow of liver damage. There was the bronze of hemochromatosis. As far as he knew, however, there was no disease that colored its victims orange. Turner's voice recalled him. In fact, he was saying, he was worried. Dr. Wooten nodded. Just what seemed to be the trouble? Turner touched his abdomen with a bright-orange hand. He had a pain down there. His abdomen had been sore off and on for over a year, but now it was more than sore. It hurt. Dr. Wooten gave an encouraging grunt, and waited. He waited for Turner to say something about his extraordinary color. But Turner had finished. He had said all he had to say. Apparently, it was only his abdomen that worried him.

Dr. Wooten stood up. He asked Turner to come along down the hall to the examining room. His color, however bizarre, could wait. A chronic abdominal pain came first. And not only that. The cause of Turner's pain was probably also the cause of his color. That seemed, at least, a reasonable assumption. They entered the examining room. Dr. Wooten switched on the light above the examination table and turned and looked at Turner. The light in his office had been an ordinary electric light, and ordinary electric light has a faintly yellow tinge. The examining room had a true-color daylight light.

But Turner's color owed nothing to tricks of light. His skin was still an unearthly golden orange. Turner stripped to the waist and got up on the table and stretched out on his back. His torso was as orange as his face. Dr. Wooten began his examination. He found the painful abdominal area, and carefully pressed. There was something there. He could feel an abnormality—a deep-seated mass about the size of an apple. It was below and behind the stomach, and he thought it might be sited at the liver. He pressed again. It wasn't the liver. It was positioned too near the center of the stomach for that. It was the pancreas.

Dr. Wooten moved away from the table. He had learned all he could from manual exploration. He waited for Turner to dress, and then led the way back to his office. He told Turner what he had found. He said he couldn't identify the mass he had felt, and he wouldn't attempt to guess. Its nature could be determined only by a series of X-ray examinations. That, he was sorry to say, would require a couple of days in the hospital. The pancreas was seated too deep to be accessible to direct X-ray examination, and an indirect examination took time and special preparation. Turner listened, and shrugged. He was willing to do whatever had to be done. Dr. Wooten swivelled around in his chair and picked up the telephone. He put in a call to the admitting office of Baptist Memorial Hospital, an affiliate of the medical school, and had a few words with the reservations clerk. He swivelled back to Turner. It was all arranged. Turner would be expected at Baptist Memorial at three o'clock that afternoon. Turner nodded, and got up to go. Dr. Wooten waved him back into his chair. There was one more thing. It was about the color of his skin. How long had it been like that? Turner looked blank. Color? What color? What was wrong with the color of his skin? Dr. Wooten hesitated. He was startled. There was no mistaking Turner's reaction. He was genuinely confused. He didn't know about his color—he really didn't know. And that was an interesting thought. It was, in fact, instructive. It clearly meant that Turner's change of color was not a sudden development. It had come on slowly, insidiously, imperceptibly. He realized that Turner was waiting, that his question had to be answered. Dr. Wooten answered it. Turner looked even blanker. He gazed at his hands, and then at Dr. Wooten. He didn't see anything unusual about his color. His skin was naturally ruddy. It always looked this way.

Dr. Wooten let it go at that. There was no point in pressing the matter any further right then. It would only worry Turner, and he was worried enough already. The matter would keep until the afternoon, until the next day, until he had a little more information to work with. He leaned back and lighted a cigarette, and changed the subject. Or seemed to. Had Turner ever met the senior associate here? That was Dr. Hughes—Dr. John D. Hughes. No? Well, in that case...Dr. Wooten reached for the telephone. Dr. Hughes's office was just next door, and he arrived a moment later. He walked into the room and glanced at Turner, and stopped—and stared. Dr. Wooten introduced them. He described the reason for Turner's visit and the mass he had found in the region of the pancreas. Dr. Hughes subdued his stare to a look of polite attention. They talked for several minutes. When Turner got up again to go, Dr. Wooten saw him to the door. He came back to his desk and sat down. Well, what did Dr. Hughes make of that? Had he ever seen or read or even heard of a man that color before? Dr. Hughes said no. And he didn't know what to think. He was completely flabbergasted. He was rather uneasy, too. That, Dr. Wooten said, made two of them.

* * *

Turner was admitted to Baptist Memorial Hospital for observation that afternoon at a few minutes after three. He was given the usual admission examination and assigned a bed in a ward. An hour or two later, Dr. Wooten, in the course of his regular hospital rounds, stopped by Turner's bed for the ritual visit of welcome and reassurance. Turner appeared to be no more than reasonably nervous, and Dr. Wooten found that satisfactory. He then turned his attention to Turner's chart and the results of the admission examination. They were, as expected, unrevealing. Turner's temperature was normal. So were his pulse rate (seventy-eight beats a minute), his respiration rate (sixteen respirations a minute), and his blood pressure (a hundred and ten systolic, eighty diastolic). The results of the urinalysis and of an electrocardiographic examination were also normal. Before resuming his rounds, Dr. Wooten satisfied himself that the really important examinations had been scheduled. These were comprehensive X-ray studies of the chest, upper gastrointestinal tract, and colon. the first two examinations were down for the following morning.

They were made at about eight o'clock. When Dr. Wooten reached the hospital on a midmorning tour, the radiologist's report was in and waiting. It more than confirmed Dr. Wooten's impression of the location of the mass. It defined its nature as well. The report read, "Lung fields are clear. Heart is normal. Barium readily traversed the esophagus and entered the stomach. In certain positions, supine projections, an apparent defect was seen on the stomach. However, this was extrinsic to the stomach. It may well represent a pseudocyst of the pancreas. No lesions of the stomach itself were demonstrated. Duodenal bulb and loop appeared normal. Stomach was emptying in a satisfactory manner." Dr. Wooten put down the report with a shiver of relief. A pancreatic cyst—even a pseudocyst—is not a trifling affliction, but he welcomed that diagnosis. The mass on Turner's pancreas just might have been a tumor. It hadn't been a likely possibility—the mass was too large and the symptoms were too mild—but it had been a possibility.

Dr. Wooten went up to Turner's ward. He told Turner what the X-ray examination had shown and what the findings meant. A cyst was a sac retaining a liquid normally excreted by the body. A pseudocyst was an empty sac—a mere dilation of space. The only known treatment of a pancreatic cyst was surgical, and surgery involving the pancreas was difficult and dangerous. Surgery was difficult because of the remote location of the pancreas, and dangerous because of the delicacy of the organs surrounding the pancreas (the stomach, the spleen, the duodenum) and the delicacy of the functions of the pancreas (the production of enzymes essential to digestion and the secretion of insulin). Fortunately, however, treatment was seldom necessary. Most cysts—particularly pseudocysts—had a way of disappearing as mysteriously as they had come. It was his belief that this was such a cyst. In that case, there was nothing much to do but be patient. And careful. Turner was to guard his belly from sudden bumps or strains. A blow or a wrench could cause a lot of trouble.

Nevertheless, Dr. Wooten went on, he wanted Turner to remain in the hospital for at least another day. There was a final X-ray of the colon to be made, and several other tests. In view of this morning's findings, the examination was, he admitted, very largely a matter of form. The cause of Turner's abdominal pain was definitely a pseudocyst of the pancreas. But prudence required an X-ray, and it would probably be done the next day. It was usual, for technical reasons, to let a day elapse between an upper-gastrointestinal study and a colon examination. Two of the other tests were indicated by the X-ray findings. One was a test for diabetes—the glucose-tolerance test. Diabetes was a possible complication of a cyst of the pancreas. Pressure from the cyst could produce diabetes by disrupting the production of insulin in the pancreatic islets of Langerhans. Such pressure could also cause another complication—a blockage of the common bile duct. The diagnostic test for that was a chemical analysis of the blood serum for the presence of the bile pigment known as bilirubin. Dr. Wooten paused. The time had come to reopen the subject that he had tactfully dropped the day before. He reopened it. It was possible, he said, that the bilirubin test might help explain the unusual color of Turner's skin. And Turner's skin *was* a most unusual color. He held up an adamant hand. No, Turner was mistaken. His color *had* changed in the past year or two. It wasn't a natural ruddiness. It was a highly unnatural orange. It was a sign that something was wrong, and he intended to find out what. That was the reason for a third test he had ordered. It was a diagnostic blood test for a condition called hemochromatosis. Hemochromatosis was a disturbance of iron metabolism that deposited iron in the skin and stained it the color of bronze. To be frank, he didn't hope for much from either of the pigmentation tests. Turner's color wasn't the bronze of hemochromatosis, and it wasn't the yellow of jaundice. the possibility of jaundice was particularly remote. The whites of Turner's eyes were still white, and that was usually where jaundice made its first appearance. But he had to carry out the tests. He had to be sure. The process of elimination was always an instructive process. And they didn't have long to wait. The results of the tests would be ready sometime that afternoon. He would be back to see Turner then.

Dr. Wooten spent the next few hours at the hospital and his office. He had other patients to see, other problems to consider, other decisions to make. But Turner remained on his mind. His first impression, like so many first impressions, had been mistaken. It now seemed practically certain that Turner's color had no connection with Turner's pancreatic cyst. They were two quite different complaints. And that returned him to the question he had asked himself when Turner walked into his office. What did an orange skin signify? What disease had the power to turn its victims orange? The answer, as before, was none. But perhaps this wasn't in the usual sense a disease. Perhaps it was a drug-induced reaction. Many chemicals in common therapeutic and diagnostic use were capable of producing conspicuous skin discolorations. Or it might be related to diet.

The question hung in Dr. Wooten's mind all day. It was still hanging there when he headed back to Turner' s ward. On the way, he picked up the results of the tests he had ordered that morning, and they did nothing to resolve it. Turner's total bilirubin level was 0.9 milligrams per hundred milliliters, or normal. The total iron-binding capacity was also normal—286 micrograms per hundred milliliters. And he didn't have diabetes. When Dr. Wooten came into the ward, he found Turner's wife at his bedside and Turner in a somewhat altered state of mind. He said he had begun to think that maybe Dr. Wooten was right about the color of his skin. There must be something peculiar about it. There had been a parade of doctors and nurses past his bed ever since early morning. Mrs. Turner looked bewildered. She hadn't noticed anything unusual about her husband's color. She hadn't thought about it—the question had never come up. But now that it had, she had to admit that he did look kind of different. He did look kind of orange. But what was the reason? What in the world could cause a thing like that? Dr. Wooten said he didn't know. The most he could say at the moment was that certain possibilities had been eliminated. He summarized the results of the three diagnostic tests. Another possible cause, he then went on to say, was drugs. Medicinal drugs. Certain medicines incorporated dyes or chemicals with pigmentary properties. Turner shook his head. Maybe so, he said, but that was out. It had been months since he had taken any kind of drug except aspirin.

Dr. Wooten was glad to believe him. Drugs had been a rather farfetched possibility. The color changes they produced were generally dramatically sudden and almost never lasting. He turned to another area—to diet. What did Turner like to eat? What, for example, did he usually have for breakfast? That was no problem, Turner said. His breakfast was almost always the same—orange juice, bacon and eggs, toast, coffee. And what about lunch? Well, that didn't change much, either. He ate a lot of vegetables—carrots, rutabagas, squash, beans, spinach, turnips, things like that. Mrs. Turner laughed. That, she said, was putting it mildly. He ate carrots the way some people eat candy. Dr. Wooten sat erect. Carrots, he was abruptly aware, were rich in carotene. So were eggs, oranges, rutabagas, squash, beans, spinach, and turnips. And carotene was a powerful yellow pigment. What, he asked Mrs. Turner, did she mean about the way her husband ate carrots? Mrs. Turner laughed again. She meant just what she said. Elmo was always eating carrots. Eating carrots and drinking tomato juice. Tomato juice was his favorite drink. And carrots were his favorite snack. He ate raw carrots all day long. He ate four or five of them a day. Why, driving down home from Alaska last week, he kept her busy just scraping and slicing and feeding him carrots. Turner gave an embarrassed grin. His wife was right. He reckoned he did eat a lot of carrots. But he had his reasons. You needed extra vitamins when you lived in Alaska. You had to make up for the long, dark winters—the lack of sunlight up there. Dr. Wooten stood up to go. What the Turners had told him was extremely interesting. He was sure, he said, that Turner's appetite for carrots was a clue to the cause of his color. It was also, as it happened, misguided. The so-called "sunshine vitamin" was vitamin D. The vitamin with which carrots and other yellow vegetables were abundantly endowed was Vitamin A.

There was a telephone just down the hall from Turner's ward. Dr. Wooten stopped and put in a call to the hospital laboratory. He arranged with the technician who took the call for a sample of Turner's blood to be tested for an abnormal concentration of carotene. Then he left the hospital and cut across the campus to the Mooney Memorial Library. He asked the librarian to let him see what she could find in the way of clinical literature of carotenemia and any related nutritional skin discolorations. He was elated by what he had learned from the Turners, but he knew that it wasn't enough. He had seen several cases of carotenemia. An excessive intake of carotene was a not uncommon condition among health-bar habitués and other amateur nutritionists. But carotene didn't color people orange. It colored them yellow. Or such had been his experience.

The librarian reported that papers on carotenemia were scarce. She had, however, found three clinical studies that looked as though they might be useful. Here was one of them. She handed Dr. Wooten a bound volume of the *Journal of the American Medical Association* for 1919, and indicated the relevant article. It was a report by two New York city investigators—Alfred F. Hess and Victor C. Myers—entitled "Carotenemia: A New Clinical Picture." Dr. Wooten knew their report, at least by reputation. It was the original study in the field. The opening descriptive paragraphs refreshed his memory and confirmed his judgment. they read:

> About a year ago one of us (A.F.H.) observed that two children in a ward containing about twenty-five infants, from a year to a year and a half in age, were developing a yellowish complexion. This coloration was not confined to the face, but involved, to a less extent, the entire body, being most evident on the palms of the hands....For a time, we were at a loss to account for this peculiar phenomenon, when our attention was directed to the fact that these two children, and only these two, were receiving a daily ration of carrots in addition to their milk and cereal. For some time we had been testing the food value of dehydrated vegetables, and when the change in color was noted, had given these babies the equivalent of 2 tablespoonfuls of fresh carrots for a period of six weeks.
>
> It seemed as if this mild jaundiced hue might well be the result of the introduction into the body of a pigment rather than the manifestation of a pathologic condition. Attention was accordingly directed to the carrots, and the same amount of this vegetable was added to the dietary of two other children of about the same age. In the one instance, after an interval of about five weeks, a yellowish tinge of the skin was noted, and about two weeks later the other baby had become somewhat yellow. There was a decided difference in the intensity of color of the four infants, indicating probably that the alteration was in part governed by individual idiosyncrasy. On omission of the carrots from the dietary, the skin gradually lost its yellow color, and in the course of some weeks regained its normal tint.

The librarian returned to Dr. Wooten's table with the other references. Both were contributions to the *New England Journal of Medicine*. One was entitled "Skin Changes of Nutritional Origin," and had been written by Harold Jeghers, an associate professor of medicine at the Boston University School of Medicine, in 1943. The other was the work of three faculty members of the Harvard Medical School—Peter Reich, Harry Shwachman, and John M. Craig—and was entitled "Lycopenemia: A Variant of Carotenemia." It had appeared in 1960. Dr. Wooten looked first at "Skin Changes of Nutritional Origin." It was a comprehensive survey, and it read, in part:

> The carotenoid group of pigments color the serum and fix themselves to the fat of the dermis and subcutaneous tissues, to which they impart the yellow tint....Edwards and Duntley showed by means of spectrophotometric analysis of skin color in human beings that carotene is present in every normal skin and is one of the five basic pigments that determine the skin color of every living person. Clinically, therefore, carotenemia refers to the presence of an excess over normal of carotene in the skin and serum....In most cases carotenemia results simply from excess use of foods rich in the carotenoid pigments. Individuals probably vary in the ease with which carotenemia develops, which is evidenced by the fact that many vegetarians do not develop it. It is said to develop more readily in those who sweat profusely. Except for the yellow color produced, it appears to be harmless, even though present for months. It eventually disappears over several weeks to months when the carotene consumption is reduced.

Dr. Wooten moved on to the third report. ("This investigation concerns a middle-aged woman whose prolonged and excessive consumption of tomato juice led to the discoloration of their skin.") He read it slowly through from beginning to end, and then turned back and reread certain passages:

> Although carotenemia due to the ingestion of foods containing a high concentration of beta carotene is a commonly described disorder, a similar condition secondary to the ingestion of tomatoes and associated with high serum levels of lycopene has not previously been reported....Lycopene is a common carotenoid pigment widely distributed through nature. It is most familiar as the red pigment of

tomatoes, but has been detected in many animals and vegetables....It is also frequently found in human serum and liver, especially when tomatoes are eaten. But lycopene is not well known medically because, unlike beta carotene, it is physiologically inert and has not been involved in any form of illness.

Dr. Wooten closed the volume. Turner was not only a heavy eater of carrots. He was also a heavy drinker of tomato juice. Carrots are rich in carotene and tomatoes are rich in lycopene. Carotene is a yellow pigment and lycopene is red. And yellow and red make orange.

Dr. Wooten completed his record of the case with a double diagnosis: pseudocyst of the pancreas and carotenemia-lycopenemia. The results of the X-ray examination of Turner's colon were normal ("Terminal ileum was visualized. No pathology was demonstrated in the colon"), and the carotene test showed a high concentration of serum carotenoids (495 micrometers per hundred milliliters, compared to a normal concentration of 50 to 350 micrometers per hundred milliliters). The diagnosis of lycopenemia was made from the clinical evidence. Turner was discharged from the hospital on December 17. His instructions were to avoid abdominal blows, carrots (and other yellow vegetables), and tomatoes in any form. Four months later, on April 16, 1961, he reported to Dr. Wooten that his skin had recovered its normal ruddiness. Two years later, in 1963, he returned again to Memphis and dropped in on Dr. Wooten for a visit. His abdominal symptoms had long since disappeared, and a comprehensive examination showed no sign of the pseudocyst.

* * *

Elmo Turner was the first recorded victim of the condition known as carotenemia-lycopenemia. He is not, however, the only one now on record. Another victim turned up in 1964. She was a woman of thirty-five, a resident of Memphis, and a patient of Dr. Wooten's. He had been treating her for a mild diabetes since 1962, and had put her at that time on the eighteen-hundred-calorie diet recommended by the American Diabetes Association. She had faithfully followed the diet, but in order to do so had eaten heavily of low-calorie vegetables, and (as she confirmed, with some surprise, when questioned) the vegetables she ate most heavily were carrots and tomatoes. She ate at least two cups of carrots and at least two whole tomatoes every day. Dr. Wooten was unaware of this until she walked into his office one October day in 1964 for her semi-annual consultation. He greeted her as calmly as he could, and asked her to sit down. He would be back in just a moment. He stepped along the hall to Dr. Hughes's office and looked in. Dr. Hughes was alone.

"Have you got a minute, John?" Dr. Wooten said.

"Sure," Dr. Hughes said. "What is it?"

"I'd like you to come into my office," Dr. Wooten said. "I'd like to show you something."

Berton Roueché

From *The Medical Detectives* (Harold Ober Associates, Inc., 1967).

Questions and Topics for Discussion and Writing

1. Outline the deductive process used by Dr. Wooten to diagnose Elmo Turner's condition. Indicate the possibilities considered by Dr. Wooten and how he confirmed or rejected each one.

2. To what extent did Dr. Wooten's diagnosis depend on his prior knowledge and training and to what extent did it depend on information gained after Elmo Turner came to see him?

3. Discuss the role of the different sources of information used by Dr. Wooten: Elmo Turner's own report, the physical examination, the laboratory tests and procedures, research in the medical literature, Mrs. Turner's comments.

Can We Know the Universe?

Carl Sagan

In this article, astronomer, science writer, and TV host, Carl Sagan discusses science as a way of thinking and scientific knowledge as a special kind of knowledge. He also considers the limits of possible human knowledge.

> Nothing is rich but the inexhaustible wealth of nature. She shows us only surfaces, but she is a million fathoms deep.
>
> *Ralph Waldo Emerson*

Science is a way of thinking much more than it is a body of knowledge. Its goal is to find out how the world works, to seek what regularities there may be, to penetrate to the connections of things—from subnuclear particles, which may be the constituents of all matter, to living organisms, the human social community, and thence to the cosmos as a whole. Our intuition is by no means an infallible guide. Our perceptions may be distorted by training and prejudice or merely because of the limitations of our sense organs, which, of course, perceive directly but a small fraction of the phenomena of the world. Even so straightforward a question as whether in the absence of friction a pound of lead falls faster than a gram of fluff was answered incorrectly by Aristotle and almost everyone else before the time of Galileo. Science is based on experiment, on a willingness to challenge old dogma, on an openness to see the universe as it really is. Accordingly, science sometimes requires courage—at the very least the courage to question the conventional wisdom.

Beyond this the main trick of science is to *really* think of something; the shape of clouds and their occasional sharp bottom edges at the same altitude everywhere in the sky; the formation of a dewdrop on a leaf; the origin of a name or a word—Shakespeare, say, or "philanthropic"; the reason for human social customs—the incest taboo, for example; how it is that a lens in sunlight can make paper burn; how a "walking stick" got to look so much like a twig; why the Moon seems to follow us as we walk; what prevents us from digging a hole down to the center of the Earth; what the definition is of "down" on a spherical Earth; how it is possible for the body to convert yesterday's lunch into today's muscle and sinew; or how far is up—does the universe go on forever, or if it does not, is there any meaning to the question of what lies on the other side? Some of these questions are pretty easy. Others, especially the last, are mysteries to which no one even today knows the answer. They are natural questions to ask. Every culture has posed such questions in one way or another. Almost always the proposed answers are in the nature of "Just So Stories," attempted explanations divorced from experiment, or even from careful comparative observations.

But the scientific cast of mind examines the world critically as if many alternative worlds might exist, as if other things might be here which are not. Then we are forced to ask why what we see is present and not something else. Why are the Sun and the Moon and the planets spheres? Why not pyramids, or cubes, or dodecahedra? Why not irregular, jumbly shapes? Why so symmetrical, worlds? If you spend any time spinning hypotheses, checking to see whether they make sense, whether they conform to what else we know, thinking of tests you can pose to substantiate or deflate your hypotheses, you will find yourself doing science. And as you come to practice this habit of thought more and more you will get better and better at it. To penetrate into the heart of the thing—even a little thing, a blade of grass, as Walt Whitman said—is to experience a kind of exhilaration that, it may be, only human beings of all the beings on this planet can feel. We are an intelligent species and the use of our intelligence quite properly gives us pleasure. In this respect the brain is like a muscle. When we think well, we feel good. Understanding is a kind of ecstasy.

But to what extent can we really know the universe around us? Sometimes this question is posed by people who hope the answer will be in the negative, who are fearful of a universe in which everything might one day be known. And sometimes we hear pronouncements from scientists who confidently state that everything worth knowing will soon be known—or even is already known—and who paint pictures of a Dionysian or Polynesian age in which the zest for intellectual discovery has withered, to be replaced by a kind of subdued languor, the lotus eaters drinking fermented coconut milk or some other mild hallucinogen. In addition to maligning both the Polynesians, who were intrepid explorers (and whose brief respite in paradise is now sadly ending), as well as the inducements to intellectual discovery provided by some hallucinogens, this contention turns out to be trivially mistaken.

Let us approach a much more modest question: not whether we can know the universe or the Milky Way Galaxy or a star or a world. Can we know, ultimately and in detail, a grain of salt? Consider one microgram of table salt, a speck just barely large enough for someone with keen eyesight to make out without a microscope. In that grain of salt there are about 10^{16} sodium and chlorine atoms. This is a 1 followed by 16 zeros, 10 million billion atoms. If we wish to know a grain of salt, we must know at least the three-dimensional positions of each of these atoms. (In fact, there is much more to be known—for example, the nature of the forces between the atoms—but we are making only a modest calculation.) Now, is this number more or less than the number of things which the brain can know?

How much *can* the brain know? There are perhaps 10^{11} neurons in the brain, the circuit elements and switches that are responsible in their electrical and chemical activity for the functioning of our minds. A typical brain neuron has perhaps a thousand little wires, called dendrites, which connect it with its fellows. If, as seems likely, every bit of information in the brain corresponds to one of these connections, the total number of things knowable by the brain is no more than 10^{14}, one hundred trillion. But this number is only one percent of the number of atoms in our speck of salt.

So in this sense the universe is intractable, astonishingly immune to any human attempt at full knowledge, We cannot on this level understand a grain of salt, much less the universe.

But let us look a little more deeply at our microgram of salt. Salt happens to be a crystal in which, except for defects in the structure of the crystal lattice, the position of every sodium and chlorine atom is predetermined. If we could shrink ourselves into this crystalline world, we would see rank upon rank of atoms in an ordered array, a regularly alternating structure—sodium, chlorine, sodium, chlorine, specifying the sheet of atoms we are standing on and all the sheets above us and below us. An absolutely pure crystal of salt could have the position of every atom specified by something like 10 bits of information.[1] This would not strain the information-carrying capacity of the brain.

If the universe had natural laws that governed its behavior to the same degree of regularity that determines a crystal of salt, then, of course, the universe would be knowable. Even if there were many such laws, each of considerable complexity, human beings might have the capability to understand them all. Even if such knowledge exceeded the information-carrying capacity of the brain, we might store the additional information outside our bodies—in books, for example, or in computer memories—and still, in some sense, know the universe.

Human beings are, understandably, highly motivated to find regularities, natural laws. The search for rules, the only possible way to understand such a vast and complex universe, is called science. The universe forces those who live in it to understand it. Those creatures who find everyday experience a muddled jumble of events with no predictability, no

[1] Chlorine is a deadly poison gas employed on European battlefields in World War I. Sodium is a corrosive metal which burns upon contact with water. Together they make a placid and unpoisonous material, table salt. Why each of these substances has the properties it does is a subject called chemistry, which requires more than 10 bits of information to understand.

regularity, are in grave peril. The universe belongs to those who, at least to some degree, have figured it out.

It is an astonishing fact that there *are* laws of nature, rules that summarize conveniently—not just qualitatively but quantitatively—how the world works. We might imagine a universe in which there are no such laws, in which the 10^{80} elementary particles that make up a universe like our own behave with utter and uncompromising abandon. To understand such a universe we would need a brain at least as massive as the universe. It seems unlikely that such a universe could have life and intelligence, because beings and brains require some degree of internal stability and order. But even if in a much more random universe there were such beings with an intelligence much greater than our own, there could not be much knowledge, passion or joy.

Fortunately for us, we live in a universe that has at least important parts that are knowable. Our common-sense experience and our evolutionary history have prepared us to understand something of the workaday world. When we go into other realms, however, common sense and ordinary intuition turn out to be highly unreliable guides. It is stunning that as we go close to the speed of light our mass increases indefinitely, we shrink toward zero thickness in the direction of motion, and time for us comes as near to stopping as we would like. Many people think that this is silly, and every week or two I get a letter from someone who complains to me about it. But it is a virtually certain consequence not just of experiment but also of Albert Einstein's brilliant analysis of space and time called the Special Theory of Relativity. It does not matter that these effects seem unreasonable to us. We are not in the habit of traveling close to the speed of light. The testimony of our common sense is suspect at high velocities.

Or consider an isolated molecule composed of two atoms shaped something like a dumbbell—a molecule of salt, it might be. Such a molecule rotates about an axis through the line connecting the two atoms. But in the world of quantum mechanics, the realm of the very small, not all orientations of our dumbbell molecule are possible. It might be that the molecule could be oriented in a horizontal position, say, or in a vertical position, but not at many angles in between. Some rotational positions are forbidden. Forbidden by what? By the laws of nature. The universe is built in such a way as to limit, or quantize, rotation. We do not experience this directly in everyday life; we would find it startling as well as awkward in sitting-up exercises, to find arms outstretched from the sides or pointed up to the skies permitted but many intermediate positions forbidden. We do not live in the world of the small, on the scale of 10^{-13} centimeters, in the realm where there are twelve zeros between the decimal place and the one. Our common-sense intuitions do not count. What does count is experiment—in this case observations from the far infrared spectra of molecules. They show molecular rotation to be quantized.

The idea that the world places restrictions on what humans might do is frustrating. Why *shouldn't* we be able to have intermediate rotational positions? Why *can't* we travel faster than the speed of light? But so far as we can tell, this is the way the universe is constructed. Such prohibitions not only press us toward a little humility; they also make the world more knowable. Every restriction corresponds to a law of nature, a regularization of the universe. The more restrictions there are on what matter and energy can do, the more knowledge human beings can attain. Whether in some sense the universe is ultimately knowable depends not only on how many natural laws there are that encompass widely divergent phenomena, but also on whether we have the openness and the intellectual capacity to understand such laws. Our formulations of the regularities of nature are surely dependent on how the brain is built, but also, and to a significant degree, on how the universe is built.

For myself, I like a universe that includes much that is unknown and, at the same time, much that is knowable. A universe in which everything is known would be static and dull, as boring as the heaven of some weak-minded theologians. A universe that is unknowable is no fit place for a thinking being. The ideal universe for us is one very much like the universe we inhabit. And I would guess that this is not really much of a coincidence.

From Carl Sagan, *Broca's Brain* (New York: Random House, 1979).

Carl Sagan

Questions and Topics for Discussion and Writing

1. Summarize Sagan's conception of what science is. What would he view as non-science?
2. In what way, according to Sagan, can we know the universe and in what way can we not know it?
3. Why are some of the implicatio science seemingly unreasonab they be reasonable to us for us them?
4. What relationship does Sagan s order in the universe and our ability to know and understand it?

We All Use the Scientific Method

Thomas Henry Huxley

Thousands of science textbooks begin with a short discussion of "the scientific method," encapsulated in a few pat sentences. These are dutifully memorized and spat forth by zillions of students on their first tests and then, whether the sentences are forgotten or not, they are at least forgotten about. But, if there *is* a method to science, what is it and how does it relate to the real world? This lecture by T. H. Huxley tries to answer these questions by characterizing scientific method in everyday life situations.

Huxley, known best as the dynamic popularizer and persuader for the theory of evolution in the nineteenth century—they called him "Darwin's Bulldog"—was a tireless writer and speaker for science in general. This lecture was one that Huxley gave to "workingmen" to help advance the cause of technical education.

The method of scientific investigation is nothing but the expression of the necessary mode of working of the human mind. It is simply the mode at which all phenomena are reasoned about, rendered precise and exact. There is no more difference, but there is just the same kind of difference, between the mental operations of a man of science and those of an ordinary person, as there is between the operations and methods of a baker or of a butcher weighing out his goods in common scales, and the operations of a chemist in performing a difficult and complex analysis by means of his balance and finely-graduated weights. It is not that the action of the scales in the one case, and the balance in the other, differ in the principles of their construction or manner of working; but the beam of one is set on an infinitely finer axis than the other, and of course turns by the addition of a much smaller weight.

You will understand this better, perhaps, if I give you some familiar example. You have all heard it repeated, I dare say, that men of science work by means of induction and deduction, and that by the help of these operations, they, in a sort of sense, wring from Nature certain other things, which are called natural laws, and causes, and that out of these, by some cunning skill of their own, they build up hypotheses and theories. And it is imagined by many, that the operations of the common mind can be by no means compared with these processes, and that they have to be acquired by a sort of special apprenticeship to the craft. To hear all these large words, you would think that the mind of a man of science must be constituted differently from that of his fellow men; but if you will not be frightened by terms, you will discover that you are quite wrong, and that all these terrible apparatus are being used by yourselves every day and every hour of your lives.

There is a well-known incident in one of Moliere's plays, where the author makes the hero express unbounded delight on being told that he had been talking prose during the whole of his life. In the same way, I trust that you will take comfort, and be delighted with yourselves, on the discovery that you have been acting on the principles of inductive and deductive philosophy during the same period. Probably there is not one here who has not in the course of the day had occasion to set in motion a complex train of reasoning, of the very same kind, though differing of course in degree, as that which a scientific man goes through in tracing the causes of natural phenomena.

A very trivial circumstance will serve to exemplify this. Suppose you go into a fruiterer's shop, wanting an apple,—you take up one, and, on biting it, you find it is sour; you look at it, and see that it is hard and green. You take up another one, and that too is hard, green, and sour. The shopman offers you a third; but, before biting it, you examine it, and find that it is hard and green, and you immediately say that you will not have it, as it must be sour, like those that you have already tried.

Nothing can be more simple than that, you think; but if you will take the trouble to analyse and trace out into its logical elements what has been done by the mind, you will be greatly surprised. In the first place, you have performed the operation of induction. You found that, in two experiences, hardness and greenness in apples went together with sourness. It was so in the first case, and it was confirmed by the second. True, it is a very small basis, but still it is enough to make an induction from; you generalise the facts, and you expect to find sourness in apples where you get hardness and greenness. You found upon that a general law, that all hard and green apples are sour; and that, so far as it goes, is a perfect induction. Well, having got your natural law in this way, when you are offered another apple which you find is hard and green, you say, "All hard and green apples are sour; this apple is hard and green, therefore this apple is sour." That train of reasoning is what logicians call a syllogism, and has all its various parts and terms—its major premise, its minor premise, and its conclusion. And, by the help of further reasoning, which, if drawn out, would have to be exhibited in two or three other syllogisms, you arrive at your final determination, "I will not have that apple." So that, you see, you have, in the first place, established a law by induction, and upon that you have founded a deduction, and reasoned out the special conclusion of the particular case. Well now, suppose, having got your law, that at some time afterwards, you are discussing the qualities of apples with a friend: you will say to him, "It is a very curious thing—but I find that all hard and green apples are sour!" Your friend says to you, "But how do you know that?" You at once reply, "Oh, because I have tried them over and over again, and have always found them to be so." Well, if we were talking science instead of common sense, we should call that an experimental verification. And, if still opposed, you go further, and say, "I have heard from the people in Somersetshire and Devonshire, where a large number of apples are grown, that they have observed the same thing. It is also found to be the case in Normandy, and in North America. In short, I find it to be the universal experience of mankind wherever attention has been directed to the subject." Whereupon, your friend, unless he is a very unreasonable man, agrees with you, and is convinced that you are quite right in the conclusion you have drawn. He believes, although perhaps he does not know he believes it, that the more extensive verifications are,—that the more frequently experiments have been made, and results of the same kind arrived at,—that the more varied the conditions under which the same results are attained, the more certain is the ultimate conclusion, and he disputes the question no further. He sees that the experiment has been tried under all sorts of conditions, as to time, place, and people, with the same result; and he says with you, therefore, that the law you have laid down must be a good one, and we must believe it.

In science we do the same thing;—the philosopher exercises precisely the same faculties, though in a much more delicate manner. In scientific inquiry it becomes a matter of duty to expose a supposed law to every possible kind of verification, and to take care, moreover, that this is done intentionally, and not left to a mere accident, as in the case of the apples. And in science, as in common life, our confidence in a law is in exact proportion to the absence of variation in the result of our experimental verifications. For instance, if you let go your grasp of an article you may have in your hand, it will immediately fall to the ground. That is a very common verification of one of the best established laws of nature—that of gravitation. The method by which men of science establish the existence of that law is exactly the same as that by which we have established the trivial proposition about the sourness of hard and green apples. But we believe it in such an extensive, thorough, and unhesitating manner because the universal experience of mankind verifies it, and we can verify it ourselves at any time; and that is the strongest possible foundation on which any natural law can rest.

So much, then, by way of proof that the method of establishing laws in science is exactly the same as that pursued in common life. Let us now turn to another matter (though really it is but another phase of the same question), and that is, the method by which, from the relations of certain phenomena, we prove that some stand in the position of causes towards the others.

I want to put the case clearly before you, and I will therefore show you what I mean by another familiar example. I will suppose that one of you, on coming down in the morning to the parlour of your house, finds that a tea-pot and some spoons which had been left in the room on the previous evening are gone,—the window is open, and you observe the mark of a dirty hand on the window-frame, and perhaps, in addition to that, you notice the impress of a hob-nailed shoe on the gravel outside. All these phenomena have struck your attention instantly, and before two seconds have passed you say, "Oh, somebody has broken open the window, entered the room, and run off with the spoons and the tea-pot!" That speech is out of your mouth in a moment. And you will probably add, "I know there has; I am quite sure of it!" You mean to say exactly what you know; but in reality you are giving expression to what is, in all essential particulars, an hypothesis. You do not *know* it at all; it is nothing but an hypothesis rapidly framed in your own mind. And it is an hypothesis founded on a long train of inductions and deductions.

What are those inductions and deductions, and how have you got at this hypothesis? You have observed, in the first place, that the window is open; but by a train of reasoning involving many inductions and deductions, you have probably arrived long before at the general law—and a very good one it is—that windows do not open of themselves; and you therefore conclude that something has opened the window. A second general law that you have arrived at in the same way is, that tea-pots and spoons do not go out of a window spontaneously, and you are satisfied that, as they are not now where you left them, they have been removed. In the third place, you look at the marks on the window-sill, and the shoe-marks outside, and you say that in all previous experience the former kind of mark has never been produced by anything else but the hand of a human being; and the same experience shows that no other animal but man at present wears shoes with hob-nails in them such as would produce the marks in the gravel. I do not know, even if we could discover any of those "missing links" that are talked about, that they would help us to any other conclusion! At any rate the law which states our present experience is strong enough for my present purpose. You next reach the conclusion, that as these kinds of marks have not been left by any other animals than men, or are liable to be formed in any other way than by a man's hand and shoe, the marks in question have been formed by a man in that way. You have, further, a general law, founded on observation and experience, and that, too, is, I am sorry to say, a very universal and unimpeachable one,—that some men are thieves; and you assume at once from all these premises—and that is what constitutes your hypothesis—that the man who made the marks outside and on the window-sill, opened the window, got into the room, and stole your tea-pot and spoons. You have now arrived at a *vera causa*;—you have assumed a cause which, it is plain, is competent to produce all the phenomena you have observed. You can explain all these phenomena only by the hypothesis of a thief. But that is a hypothetical conclusion, of the justice of which you have no absolute proof at all; it is only rendered highly probable by a series of inductive and deductive reasonings.

I suppose your first action, assuming that you are a man of ordinary common sense, and that you have established this hypothesis to your own satisfaction, will very likely be to go off for the police, and set them on the track of the burglar, with the view to the recovery of your property. But just as you are starting with this object, some person comes in, and on learning what you are about, says, "My good friend, you are going on a great deal too fast. How do you know that the man who really made the marks took the spoons? It might have been a monkey that took them, and the man may have merely looked in afterwards." You would probably reply, "Well, that is all very well, but you see it is contrary to all experience of the way tea-pots and spoons are abstracted; so that, at any rate, your hypothesis is less probable than mine." While you are talking the thing over in this way, another friend arrives, one of that good kind of people that I was talking of a little while ago. And he might say, "Oh, my dear sir, you are certainly going on a great deal too fast. You are most presumptuous. You admit that all these occurrences took place when you were fast asleep, at a time when you could not possibly have known anything about what was taking place. How do you know that the laws of Nature are not suspended

Thomas Henry Huxley

during the night? It may be that there has been some kind of supernatural interference in this case." In point of fact, he declares that your hypothesis is one of which you cannot at all demonstrate the truth, and that you are by no means sure that the laws of Nature are the same when you are asleep as when you are awake.

Well, now, you cannot at the moment answer that kind of reasoning. You feel that your worthy friend has you somewhat at a disadvantage. You will feel perfectly convinced in your own mind, however, that you are quite right, and you say to him, "My good friend, I can only be guided by the natural probabilities of the case, and if you will be kind enough to stand aside and permit me to pass, I will go and fetch the police." Well, we will suppose that your journey is successful, and that by good luck you meet with a policeman; that eventually the burglar is found with your property on his person, and the marks correspond to his hand and to his boots. Probably any jury would consider those facts a very good experimental verification of your hypothesis, touching the cause of the abnormal phenomena observed in your parlour, and would act accordingly.

Now, in this suppositious case, I have taken phenomena of a very common kind, in order that you might see what are the different steps in an ordinary process of reasoning, if you will only take the trouble to analyse it carefully. All the operations I have described, you will see, are involved in the mind of any man of sense in leading him to a conclusion as to the course he should take in order to make good a robbery and punish the offender. I say that you are led, in that case, to your conclusion by exactly the same train of reasoning as that which a man of science pursues when he is endeavouring to discover the origin and laws of the most occult phenomena. The process is, and always must be, the same; and precisely the same mode of reasoning was employed by Newton and Laplace in their endeavours to discover and define the causes of the movements of the heavenly bodies, as you, with your own common sense, would employ to detect a burglar. The only difference is, that the nature of the inquiry being more abstruse, every step has to be most carefully watched, so that there may not be a single crack or flaw in your hypothesis. A flaw or crack in many of the hypotheses of daily life may be of little or no moment as affecting the general correctness of the conclusions at which we may arrive; but, in a scientific inquiry, a fallacy, great or small, is always of importance, and is sure to be in the long run constantly productive of mischievous, if not fatal results.

Do not allow yourselves to be misled by the common notion that an hypothesis is untrustworthy simply because it is an hypothesis. It is often urged, in respect to some scientific conclusion, that, after all, it is only an hypothesis. But what more have we to guide us in nine-tenths of the most important affairs of daily life than hypotheses, and often very ill-based ones? So that in science, where the evidence of an hypothesis is subjected to the most rigid examination, we may rightly pursue the same course. You may have hypotheses and hypotheses. A man may say, if he likes, that the moon is made of green cheese: that is an hypothesis. But another man, who has devoted a great deal of time and attention to the subject, and availed himself of the most powerful telescopes and the results of the observations of others, declares that in his opinion it is probably composed of materials very similar to those of which our own earth is made up: and that is also only an hypothesis. But I need not tell you that there is an enormous difference in the value of the two hypotheses. That one which is based on sound scientific knowledge is sure to have a corresponding value; and that which is a mere hasty random guess is likely to have but little value. Every great step in our progress in discovering causes has been made in exactly the same way as that which I have detailed to you. A person observing the occurrence of certain facts and phenomena asks, naturally enough, what process, what kind of operation known to occur in Nature applied to the particular case, will unravel and explain the mystery? Hence you have the scientific hypothesis; and its value will be proportionate to the care and completeness with which its basis had been tested and verified. It is in these matters as in the commonest affairs of practical life: the guess of the fool will be folly, while the guess of the wise man will contain wisdom. In all cases, you see that the value of the result depends on the patience and faithfulness with which the investigator applies to his hypothesis every possible kind of verification.

Thomas Henry Huxley

Questions and Topics for Discussion and Writing

1. Describe the process of induction and the formation of a hypothesis illustrated by the hard, green apples.
2. Discuss the objections raised against the conclusion that a thief has stolen the spoons and the teapot. What is the role of past experience in the evaluation of the hypothesis versus the objections to it?
3. Why does Huxley say that the two hypotheses about the composition of the moon have great differences in value? What makes the hypothesis that the moon is similar to the earth more valuable that the hypothesis that it is made of green cheese?

* Scientific Inquiry

Carl G. Hempel

If there is a methodology to scientific work, then there must be a logic to it, a logic that can be examined and analysed by philosophers. Philosophers who devote themselves to the analysis of the logic and methodology of science call themselves philosophers of science. One of the most prominent of these is Professor Carl G. Hempel of Princeton University. This excerpt from his book, *Philosophy of Natural Science,* explores the logic of forming and testing hypotheses. Hempel illustrates his analysis with a truly fascinating case study from the history of medicine.

A Case History as an Example

As a simple illustration of some important aspects of scientific inquiry let us consider Semmelweis' work on childbed fever. Ignaz Semmelweis, a physician of Hungarian birth, did this work during the years from 1844 to 1848 at the Vienna General Hospital. As a member of the medical staff of the First Maternity Division in the hospital, Semmelweis was distressed to find that a large proportion of the women who were delivered of their babies in that division contracted a serious and often fatal illness known as puerperal fever or childbed fever. In 1844, as many as 260 out of 3,157 mothers in the First Division, or 8.2 per cent, died of the disease; for 1845, the death rate was 6.8 per cent, and for 1846, it was 11.4 per cent. These figures were all the more alarming because in the adjacent Second Maternity Division of the same hospital, which accommodated almost as many women as the First, the death toll from childbed fever was much lower: 2.3, 2.0, and 2.7 per cent for the same years. In a book that he wrote later on the causation and the prevention of childbed fever, Semmelweis describes his efforts to resolve the dreadful puzzle.[1]

He began by considering various explanations that were current at the time; some of these he rejected out of hand as incompatible with well-established facts; others he subjected to specific tests.

One widely accepted view attributed the ravages of puerperal fever to "epidemic influences," which were vaguely described as "atmospheric-cosmic-telluric changes" spreading over whole districts and causing childbed fever in women in confinement. But how, Semmelweis reasons, could such influences have plagued the First Division for years and yet spared the Second? And how could this view he reconciled with the fact that while the fever was raging in the hospital, hardly a case occurred in the city of Vienna or in its surroundings: a genuine epidemic, such as cholera, would not be so selective. Finally, Semmelweis notes that some of the women admitted to the First Division, living far from the hospital, had been overcome by labor on their way and had given birth in the street: yet despite these adverse conditions, the death rate from childbed fever among these cases of "street birth" was lower than the average for the First Division.

On another view, overcrowding was a cause of mortality in the First Division. But Semmelweis points out that in fact the crowding was heavier in the Second Division, partly as a result of the desperate efforts of patients to avoid assignment to the

[1] The story of Semmelweis' work and of the difficulties he encountered forms a fascinating page in the history of medicine. A detailed account, which includes translations and paraphrases of large portions of Semmelweis' writings, is given in W. J. Sinclair, *Semmelweis: His Life and His Doctrine* (Manchester, England: Manchester University Press, 1909). Brief quoted phrases in this chapter are taken from this work. The highlights of Semmelweis' career are recounted in the first chapter of P. de Kruif, *Men Against Death* (New York: Harcourt, Brace & World, Inc., 1932).

notorious First Division. He also rejects two similar conjectures that were current, by noting that there were no differences between the two Divisions in regard to diet or general care of the patients.

In 1846, a commission that had been appointed to investigate the matter attributed the prevalence of illness in the First Division to injuries resulting from rough examination by the medical students, all of whom received their obstetrical training in the First Division. Semmelweis notes in refutation of this view that (a) the injuries resulting naturally from the process of birth are much more extensive than those that might be caused by rough examination; (b) the midwives who received their training in the Second Division examined their patients in much the same manner but without the same ill effects; (c) when, in response to the commission's report, the number of medical students was halved and their examinations of the women were reduced to a minimum, the mortality, after a brief decline, rose to higher levels than ever before.

Various psychological explanations were attempted. One of them noted that the First Division was so arranged that a priest bearing the last sacrament to a dying woman had to pass through five wards before reaching the sickroom beyond: the appearance of the priest, preceded by an attendant ringing a bell, was held to have a terrifying and debilitating effect upon the patients in the wards and thus to make them more likely victims of childbed fever. In the Second Division, this adverse factor was absent, since the priest had direct access to the sickroom. Semmelweis decided to test this conjecture. He persuaded the priest to come by a roundabout route and without ringing of the bell, in order to reach the sick chamber silently and unobserved. But the mortality in the First Division did not decrease.

A new idea was suggested to Semmelweis by the observation that in the First Division the women were delivered lying on their backs; in the Second Division, on their sides. Though he thought it unlikely, he decided "like a drowning man clutching at a straw," to test whether this difference in procedure was significant. He introduced the use of the lateral position in the First Division, but again, the mortality remained unaffected.

At last, early in 1847, an accident gave Semmelweis the decisive clue for his solution of the problem. A colleague of his, Kolletschka, received a puncture wound in the finger, from the scalpel of a student with whom he was performing an autopsy, and died after an agonizing illness during which he displayed the same symptoms that Semmelweis had observed in the victims of childbed fever. Although the role of microorganisms in such infections had not yet been recognized at the time, Semmelweis realized that "cadaveric matter" which the student's scalpel had introduced into Kolletschka's blood stream had caused his colleague's fatal illness. And the similarities between the course of Kolletschka's disease and that of the women in his clinic led Semmelweis to the conclusion that his patients had died of the same kind of blood poisoning: he, his colleagues, and the medical students had been the carriers of the infectious material, for he and his associates used to come to the wards directly from performing dissections in the autopsy room, and examine the women in labor after only superficially washing their hands, which often retained a characteristic foul odor.

Again, Semmelweis put his idea to a test. He reasoned that if he were right, then childbed fever could be prevented by chemically destroying the infectious material adhering to the hands. He therefore issued an order requiring all medical students to wash their bands in a solution of chlorinated lime before making an examination. The mortality from childbed fever promptly began to decrease, and for the year 1848 it fell to 1.27 per cent in the First Division, compared to 1.33 in the Second.

In further support of his idea, or of his *hypothesis*, as we will also say, Semmelweis notes that it accounts for the fact that the mortality in the Second Division consistently was so much lower: the patients there were attended by midwives, whose training did not include anatomical instruction by dissection of cadavers.

The hypothesis also explained the lower mortality among "street births": women who arrived with babies in arms were rarely examined after admission and thus had a better chance of escaping infection.

Similarly, the hypothesis accounted for the fact that the victims of childbed fever among the newborn babies were all among those whose mothers had contracted the disease during labor; for then the infection could be transmitted to the baby before birth, through the common bloodstream of mother and child, whereas this was impossible when the mother remained healthy.

Further clinical experiences soon led Semmelweis to broaden his hypothesis. On one occasion, for example, he and his associates, having carefully disinfected their hands, examined first a woman in labor who was suffering from a festering cervical cancer; then they proceeded to examine twelve other women in the same room, after only routine washing without renewed disinfection. Eleven of the twelve patients died of puerperal fever. Semmelweis concluded that childbed fever can be caused not only by cadaveric material, but also by "putrid matter derived from living organisms."

Basic Steps in Testing a Hypothesis

We have seen how, in his search for the cause of childbed fever, Semmelweis examined various hypotheses that had been suggested as possible answers. How such hypotheses are arrived at in the first place is an intriguing question which we will consider later. First, however, let us examine how a hypothesis, once proposed, is tested.

Sometimes, the procedure is quite direct. Consider the conjectures that differences in crowding, or in diet, or in general care account for the difference in mortality between the two divisions. As Semmelweis points out, these conflict with readily observable facts. There are no such differences between the divisions; the hypotheses are therefore rejected as false.

But usually the test will be less simple and straightforward. Take the hypothesis attributing the high mortality in the First Division to the dread evoked by the appearance of the priest with his attendant. The intensity of that dread, and especially its effect upon childbed fever, are not as directly ascertainable as are differences in crowding or in diet, and Semmelweis uses an indirect method of testing. He asks himself: Are there any readily observable effects that should occur if the hypothesis were true? And he reasons: *If* the hypothesis were true, *then* an appropriate change in the priest's procedure should be followed by a decline in fatalities. He checks this implication by a simple experiment and finds it false, and he therefore rejects the hypothesis.

Similarly, to test his conjecture about the position of the women during delivery, he reasons: *If* this conjecture should be true, *then* adoption of the lateral position in the First Division will reduce the mortality. Again, the implication is shown false by his experiment, and the conjecture is discarded.

In the last two cases, the test is based on an argument to the effect that *if* the contemplated hypothesis, say H, is true, then certain observable events (e.g., decline in mortality) should occur under specified circumstances (e.g., if the priest refrains from walking through the wards, or if the women are delivered in lateral position); or briefly, if H is true, then so is I, where I is a statement describing the observable occurrences to be expected. For convenience, let us say that I is inferred from, or implied by, H; and let us call I a *test implication of the hypothesis H.*

In our last two examples, experiments show the test implication to be false, and the hypothesis is accordingly rejected. The reasoning that leads to the rejection may be schematized as follows:

If H is true, then so is I.
But (as the evidence shows) I is not true.
(*a*) ________________
H is not true.

Any argument of this form, called *modus tollens* in logic, is deductively valid; that is, if its premisses (the sentences above the horizontal line) are true, then its conclusion (the sentence below the horizontal line) is unfailingly true as well. Hence, if the premisses of *(a)* are properly established, the hypothesis H that is being tested must indeed be rejected.

Next, let us consider the case where observation or experiment bears out the test implication I. From his hypothesis that childbed fever is blood poisoning produced by cadaveric matter, Semmelweis infers that suitable antiseptic measures will reduce fatalities from the disease. This time, experiment shows the test implication to be true. But this favorable outcome does not conclusively prove the hypothesis true, for the underlying argument would have the form

If H is true, then so is I.
(b) *(As the evidence shows) I is true.*
____
H is true.

And this mode of reasoning, which is referred to as the *fallacy of affirming the consequent*, is deductively invalid, that is, its conclusion may be false even if its premisses are true. This is in fact illustrated by Semmelweis' own experience. The initial version of his account of childbed fever as a form of blood poisoning presented infection with cadaveric matter essentially as the one and only source of the disease; and he was right in reasoning that if this hypothesis should be true, then destruction of cadaveric particles by antiseptic washing should reduce the mortality. Furthermore, his experiment did show the test implication to be true. Hence, in this case, the premisses of *(b)* were both true. Yet, his hypothesis was false, for as he later discovered, putrid material from living organisms, too, could produce childbed fever.

Thus, the favorable outcome of a test, i.e., the fact that a test implication inferred from a hypothesis is found to be true, does not prove the hypothesis to be true. Even if many implications of a hypothesis have been borne out by careful tests, the hypothesis may still be false. The following argument still commits the fallacy of affirming the consequent:

If H is true, then so are $I_1, I_2, ..., I_n$.
(c) *(As the evidence shows* $I_1, I_2, ..., I_n$ *are all true.*
____
H is true.

This, too, can be illustrated by reference to Semmelweis' final hypothesis in its first version. As we noted earlier, his hypothesis also yields the test implications that among cases of street births admitted to the First Division, mortality from puerperal fever should be below the average for the Division, and that infants of mothers who escape the illness do not contract childbed fever; and these implications, too, were borne out by the evidence—even though the first version of the final hypothesis was false.

But the observation that a favorable outcome of however many tests does not afford conclusive proof for a hypothesis should not lead us to think that if we have subjected a hypothesis to a number of tests and all of them have had a favorable outcome, we are no better off than if we had not tested the hypothesis at all. For each of our tests might conceivably have had an unfavorable outcome and might have led to the rejection of the hypothesis. A set of favorable results obtained by testing different test implications, $I_1, I_2, ..., I_n$ of a hypothesis, shows that as far as these particular implications are concerned, the hypothesis has been borne out; and while this result does not afford a complete proof of the hypothesis, it provides at least some support, some partial corroboration or confirmation for it. The extent of this support will depend on various aspects of the hypothesis and of the test data ...

Let us now consider another example,[1] which will also bring to our attention some further aspects of scientific inquiry.

As was known at Galileo's time, and probably much earlier, a simple suction pump, which draws water from a well by means of a piston that can be raised in the pump barrel, will lift water no higher than about 34 feet above the surface of the well. Galileo was intrigued by this limitation and suggested an explanation for it, which was, however, unsound. After Galileo's death, his pupil Torricelli advanced a new answer. He argued that the earth is surrounded by a sea of air, which, by reason of its weight exerts pressure upon the surface below, and that this pressure upon the surface of the well forces water up the pump barrel when the piston is raised. The maximum length of 34 feet for the water column in the barrel thus reflects simply the total pressure of the atmosphere upon the surface of the well.

It is evidently impossible to determine by direct inspection or observation whether this account is correct, and Torricelli tested it indirectly. He reasoned that *if* his conjecture were true, *then* the pressure of the atmosphere should also be capable

[1] The reader will find a fuller account of this example in Chap. 4 of J. B. Conant's fascinating book, *Science and Common Sense* (New Haven: Yale University Press, 1951). A letter by Torricelli setting forth his hypothesis and his test of it, and an eyewitness report on the Puy-de-Dôme experiment are reprinted in W. F. Magie, *A Source Book in Physics* (Cambridge: Harvard University Press, 1963), pp. 70–75.

of supporting a proportionately shorter column of mercury; indeed, since the specific gravity of mercury is about 14 times that of water, the length of the mercury column should be about $\frac{34}{14}$ feet, or slightly less than 2 ½ feet. He checked this test implication by means of an ingeniously simple device, which was, in effect, the mercury barometer. The well of water is replaced by an open vessel, containing mercury; the barrel of the suction pump is replaced by a glass tube sealed off at one end. The tube is completely filled with mercury and closed by placing the thumb tightly over the open end. It is then inverted, the open end is submerged in the mercury well, and the thumb is withdrawn; whereupon the mercury column in the tube drops until its length is about 30 inches—just as predicted by Torricelli's hypothesis.

A further test implication of that hypothesis was noted by Pascal, who reasoned that if the mercury in Torricelli's barometer is counterbalanced by the pressure of the air above the open mercury well, then its length should decrease with increasing altitude, since the weight of the air overhead becomes smaller. At Pascal's request, this implication was checked by his brother-in-law, Périer, who measured the length of the mercury column in the Torricelli barometer at the foot of the Puy-de-Dôme, a mountain some 4,800 feet high, and then carefully carried the apparatus to the top and repeated the measurement there while a control barometer was left at the bottom under the supervision of an assistant. Périer found the mercury column at the top of the mountain more than three inches shorter than at the bottom, whereas the length of the column in the control barometer had remained unchanged throughout the day.

The Role of Induction in Scientific Inquiry

We have considered some scientific investigations in which a problem was tackled by proposing tentative answers in the form of hypotheses that were then tested by deriving from them suitable test implications and checking these by observation or experiment.

But how are suitable hypotheses arrived at in the first place?

It is sometimes held that they are inferred from antecedently collected data by means of a procedure called *inductive inference*, as contradistinguished from deductive inference, from which it differs in important respects.

In a deductively valid argument, the conclusion is related to the premisses in such a way that if the premisses are true then the conclusion cannot fail to be true as well. This requirement is satisfied, for example, by any argument of the following general form:

If p, then q.
It is not the case that q.

It is not the case that p.

Brief reflection shows that no matter what particular statements may stand at the places marked by the letters "p" and "q," the conclusion will certainly be true if the premisses are. In fact, our schema represents the argument form called *modus tollens*, to which we referred earlier.

Another type of deductively valid inference is illustrated by this example:

Any sodium salt, when put into the flame of a Bunsen burner, turns the flame yellow.
This piece of rock salt is a sodium salt.

This piece of rock salt, when put into the flame of a Bunsen burner, will turn the flame yellow.

Arguments of the latter kind are often said to lead from the general (here, the premiss about all sodium salts) to the particular (a conclusion about the particular piece of rock salt). Inductive inferences, by contrast, are sometimes described as leading from premisses about particular cases to a conclusion that has the character of a general law or principle. For example, from premisses to the effect that each of the particular samples of various sodium salts that have so far been subjected to the Bunsen flame test did turn the flame yellow, inductive inference supposedly leads to the general conclusion that all sodium salts, when put into the flame of a Bunsen burner, turn the flame yellow. But in this case, the truth of the premisses obviously does *not* guarantee the truth of the conclusion; for even if it is the case that all samples of sodium salts examined so far did turn the Bunsen flame yellow, it remains quite possible that new kinds of sodium salt might yet be found that do not conform to this generalization. Indeed, even some kinds of sodium salt that have already been tested with positive result might conceivably fail to satisfy the generalization under special physical conditions (such as very strong magnetic fields or the like) in which they have not yet been examined. For this reason, the premisses of an inductive inference are often said to imply the conclusion only with more or less high probability, whereas the premisses of a deductive inference imply the conclusion with certainty.

The idea that in scientific inquiry, inductive inference from antecedently collected data leads to appropriate general principles is clearly embodied in the following account of how a scientist would ideally proceed:

> If we try to imagine how a mind of superhuman power and reach, but normal so far as the logical processes of its thought are concerned, ... would use the scientific method, the process would be as follows: First, all facts would be observed and recorded, *without selection* or *a priori* guess as to their relative importance. Secondly, the observed and recorded facts would be analyzed, compared, and classified, *without hypothesis or postulates* other than those necessarily involved in the logic of thought. Third, from this analysis of the facts generalizations would be inductively drawn as to the relations, classificatory or causal, between them. Fourth, further research would be deductive as well as inductive, employing inferences from previously established generalizations.[1]

This passage distinguishes four stages in an ideal scientific inquiry: (1) observation and recording of all facts, (2) analysis and classification of these facts, (3) inductive derivation of generalizations from them, and (4) further testing of the generalizations. The first two of these stages are specifically assumed not to make use of any guesses or hypotheses as to how the observed facts might be interconnected; this restriction seems to have been imposed in the belief that such preconceived ideas would introduce a bias and would jeopardize the scientific objectivity of the investigation.

But the view expressed in the quoted passage—I will call it *the narrow inductivist conception of scientific inquiry*—is untenable, for several reasons. A brief survey of these can serve to amplify and to supplement our earlier remarks on scientific procedure.

First, a scientific investigation as here envisaged could never get off the ground. Even its first phase could never be carried out, for a collection of *all* the facts would have to await the end of the world, so to speak; and even all the facts *up to now* cannot be collected, since there are an infinite number and variety of them. Are we to examine, for, example, all the grains of sand in all the deserts and on all the beaches, and are we to record their shapes, their weights, their chemical composition, their distances from each other, their constantly changing temperature, and their equally changing distance from the center of the moon? Are we to record the floating thoughts that cross our minds in the tedious process? The shapes of the clouds overhead, the changing color of the sky? The construction and the trade name of our writing equipment? Our own life histories and those of our fellow investigators? All these and untold other things, are, after all, among "all the facts up to now."

Perhaps, then, all that should be required in the first phase is that all the *relevant* facts be collected. But relevant to what? Though the author does not mention this, let us suppose that the inquiry is concerned with a specified *problem*. Should we not then begin by collecting all the facts—or better, all available data—relevant to that problem? This notion still makes no clear sense. Semmelweis sought to solve one specific problem, yet he collected quite different kinds of data at different stages of his inquiry. And rightly so; for what particular sorts of data it is reasonable to collect is not determined by the problem under study, but by a tentative answer to it that the investigator entertains in the form of a conjecture or hypothesis. Given the conjecture that mortality from childbed fever was increased by the terrifying appearance of the priest and his attendant with the death bell, it was relevant to collect data on the consequences of having the priest change his routine; but it would have been totally irrelevant to check what would happen if doctors and students disinfected their hands before examining their patients. With respect to Semmelweis' eventual contamination hypothesis, data of the latter kind were

[1] A. B. Wolfe, "Functional Economics," in *The Trend of Economics*, ed. R. G. Tugwell, 450 (New York: Alfred A. Knopf, 1924) [italics are quoted].

clearly relevant, and those of the former kind totally irrelevant.

Empirical "facts" or findings, therefore, can be qualified as logically relevant or irrelevant only in reference to a given hypothesis, but not in reference to a given problem.

Suppose now that a hypothesis *H* bas been advanced as a tentative answer to a research problem: what kinds of data would be relevant to *H*? Our earlier examples suggest an answer: A finding is relevant to *H* if either its occurrence or its nonoccurrence can be inferred from *H*. Take Torricelli's hypothesis, for example. As we saw, Pascal inferred from it that the mercury column in a barometer should grow shorter if the barometer were carried up a mountain. Therefore, any finding to the effect that this did indeed happen in a particular case is relevant to the hypotheses; but so would be the finding that the length of the mercury column had remained unchanged or that it had decreased and then increased during the ascent, for such findings would refute Pascal's test implication and would thus disconfirm Torricelli's hypothesis. Data of the former kind may be called positively, or favorably, relevant to the hypothesis; those of the latter kind negatively, or unfavorably, relevant.

In sum, the maxim that data should be gathered without guidance by antecedent hypotheses about the connections among the facts under study is self-defeating, and it is certainly not followed in scientific inquiry. On the contrary, tentative hypotheses are needed to give direction to a scientific investigation. Such hypotheses determine, among other things, what data should be collected at a given point in a scientific investigation.

It is of interest to note that social scientists trying to check a hypothesis by reference to the vast store of facts recorded by the U.S. Bureau of the Census, or by other data-gathering organizations, sometimes find to their disappointment that the values of some variable that plays a central role in the hypothesis have nowhere been systematically recorded. This remark is not, of course, intended as a criticism of data gathering: those engaged in the process no doubt try to select facts that might prove relevant to future hypotheses; the observation is simply meant to illustrate the impossibility of collecting "all the relevant data" without knowledge of the hypotheses to which the data are to have relevance.

The second stage envisaged in our quoted passage is open to similar criticism. A set of empirical "facts" can be analyzed and classified in many different ways, most of which will be unilluminating for the purposes of a given inquiry. Semmelweis could have classified the women in the maternity wards according to criteria such as age, place of residence, marital status, dietary habits, and so forth; but information on these would have provided no clue to a patient's prospects of becoming a victim of childbed fever. What Semmelweis sought were criteria that would be significantly connected with those prospects; and for this purpose, as he eventually found, it was illuminating to single out those women who were attended by medical personnel with contaminated hands; for it was with this characteristic, or with the corresponding class of patients, that high mortality from childbed fever was associated.

Thus, if a particular way of analyzing and classifying empirical findings is to lead to an explanation of the phenomena concerned, then it must be based on hypotheses about how those phenomena are connected; without such hypotheses, analysis and classification are blind.

Our critical reflections on the first two stages of inquiry as envisaged in the quoted passage also undercut the notion that hypotheses are introduced only in the third stage, by inductive inference from antecedently collected data. But some further remarks on the subject should be added here.

Induction is sometimes conceived as a method that leads, by means of mechanically applicable rules, from observed facts to corresponding general principles. In this case, the rules of inductive inference would provide effective canons of scientific discovery; induction would be a mechanical procedure analogous to the familiar routine for the multiplication of integers, which leads, in a finite number of predetermined and mechanically performable steps, to the corresponding product. Actually, however, no such general and mechanical induction procedure is available at present; otherwise, the much studied problem of the causation of cancer, for example, would hardly have remained unsolved to this day. Nor can the discovery of such a procedure ever be expected. For—to mention one reason—scientific hypotheses and theories are usually couched in terms that do not occur at all in the description of the empirical findings on which they rest, and which they serve to explain. For example, theories about the atomic and subatomic structure of matter contain terms such as "atom," "electron," "proton," "neutron," "psi-function," etc.; yet they are based on laboratory findings about the spectra of various gases, tracks in cloud and bubble chambers, quantitative aspects of chemical reactions, and so forth—all of which can be described without the use of those "theoretical terms." Induction rules of the kind here envisaged would therefore have to provide a mechanical routine for constructing, on the basis of the given data, a hypothesis or theory stated in terms of some quite novel concepts, which are nowhere used in the description of the data themselves. Surely, no general mechanical rule of procedure can be expected to achieve this. Could there be a general rule, for example, which, when applied to the data available to Galileo concerning the limited effectiveness of suction pumps, would, by a mechanical routine, produce a hypothesis based on the concept of a sea of air?

To be sure, mechanical procedures for inductively "inferring" a hypothesis on the basis of given data may be specifiable for situations of special, and relatively simple, kinds. For example, if the length of a copper rod bas been measured at several different temperatures, the resulting pairs of associated values for temperature and length may be represented by points in a plane coordinate system, and a curve may be drawn through them in accordance with some particular rule of curve fitting. The curve then graphically represents a general quantitative hypothesis that expresses the length of the rod as a specific function of its temperature. But note that this hypothesis contains no novel terms; it is expressible in terms of the concepts of temperature and length, which are used also in describing the data. Moreover, the choice of "associated" values of temperature and length as data already presupposes a guiding hypothesis; namely that with each value of the temperature, exactly one value of the length of the copper rod is associated, so that its length is indeed a function of its temperature alone. The mechanical curve-fitting routine then serves only to select a particular function as the appropriate one. This point is important; for suppose that instead of a copper rod, we examine a body of nitrogen gas enclosed in a cylindrical container with a movable piston as a lid, and that we measure its volume at several different temperatures. If we were to use this procedure in an effort to obtain from our data a *general* hypothesis representing the volume of the gas as a function of its temperature, we would fail, because the volume of a gas is a function both of its temperature and of the pressure exerted upon it, so that at the same temperature, the given gas may assume different volumes.

Thus, even in these simple cases, the mechanical procedures for the construction of a hypothesis do only part of the job, for they presuppose an antecedent, less specific hypothesis (i.e., that a certain physical variable is a function of one single other variable), which is not obtainable by the same procedure.

There are, then, no generally applicable "rules of induction," by which hypotheses or theories can be mechanically derived or inferred from empirical data. The transition from data to theory requires creative imagination. Scientific hypotheses and theories are not *derived* from observed facts, but *invented* in order to account for them. They constitute guesses at the connections that might obtain between the phenomena under study, at uniformities and patterns that might underlie their occurrence. "Happy guesses" of this kind require great ingenuity, especially if they involve a radical departure from current modes of scientific thinking, as did, for example, the theory of relativity and quantum theory. The inventive effort required in scientific research will benefit from a thorough familiarity with current knowledge in the field. A complete novice will hardly make an important scientific discovery, for the ideas that may occur to him are likely to duplicate what has been tried before or to run afoul of well-established facts or theories of which he is not aware ...

Scientific knowledge, as we have seen, is not arrived at by applying some inductive inference procedure to antecedently collected data, but rather by what is often called "the method of hypothesis," i.e. by inventing hypotheses as tentative answers to a problem under study, and then subjecting these to empirical test. It will be part of such test to see whether the hypothesis is borne out by whatever relevant findings may have been gathered before its formulation; an acceptable hypothesis will have to fit the available relevant data. Another part of the test will consist in deriving new test implications from the hypothesis and checking these by suitable observations or experiments. As we noted earlier, even extensive testing with entirely favorable results does not establish a hypothesis conclusively, but provides only more or less strong support for it. Hence, while scientific inquiry is certainly not inductive in the narrow sense we have examined in some detail, it may be said to be *inductive in a wider sense*, inasmuch as it involves the acceptance of hypotheses on the basis of data that afford no deductively conclusive evidence for it, but lend it only more or less strong "inductive support" or confirmation. And any "rules of induction" will have to be conceived, in analogy to the rules of deduction, as canons of validation rather than of discovery. Far from generating a hypothesis that accounts for given empirical findings, such rules will presuppose that both the empirical data forming the "premisses" of the "inductive argument" and a tentative hypothesis forming its "conclusion" are *given*. The rules of induction would then state criteria for the soundness of the argument. According to some theories of induction, the rules would determine the strength of the support that the data lend to the hypothesis, and they might express such support in terms of probabilities.

From Carl G. Hempel, *Philosophy of Natural Science* (Englewood Cliffs, NJ: Prentice-Hall, 1966).

Questions and Topics for Discussion and Writing

1. Why did Hempel choose to illustrate the process of scientific inquiry by using the case of Semmelweis's discovery of the causes of childbed fever? Show how each of the steps of Semmelweis's investigation illustrate different kinds of reasoning.
2. What is the role of testing in this process? How are effective tests of Semmelweis's hypotheses constructed? Is this a straightforward mechanical process or does it require an inventiveness comparable to making the hypothesis in the first place?
3. Why is the "ideal scientific inquiry" sketched by Hempel an impossibility? Discuss how Galileo might have attempted to explain the limitation on the effectiveness of a suction pump using the method of "ideal scientific inquiry."

The Discovery of Radium

Marie Curie

The following short article is from an address by one of the world's greatest experimental physicists, Marie Curie, to an audience at Vassar College in Massachusetts acknowledging a gift by American women to her of a gram of radium for use in her research. Radium was (and is) extremely expensive and difficult to produce.

Marie Curie (1867–1934) was the first person ever to receive a second Nobel Prize. The first, the Nobel Prize in physics in 1903, she shared with her husband, Pierre Curie, and Henri Becquerel. The second, the Nobel Prize in chemistry in 1911, she received alone. Her life was one of tremendous hardships and strenuous work. After Pierre Curie's tragic death in a traffic accident, Marie Curie took over his teaching position and thereby became the first woman professor at the Sorbonne in its 650-year history.

In this article she describes the general course of her work on radiation.

I could tell you many things about radium and radioactivity and it would take a long time. But as we cannot do that, I shall give you only a short account of my early work about radium. Radium is no more a baby; it is more than twenty years old, but the conditions of the discovery were somewhat peculiar, and so it is always of interest to remember them and to explain them.

We must go back to the year 1897. Professor Curie and I worked at that time in the laboratory of the School of Physics and Chemistry where Professor Curie held his lectures. I was engaged in some work on uranium rays which had been discovered two years before by Professor Becquerel. I shall tell you how these uranium rays may be detected. If you take a photographic plate and wrap it in black paper and then on this plate, protected from ordinary light, put some uranium salt and leave it a day, and the next day the plate is developed, you notice on the plate a black spot at the place where the uranium salt was. This spot has been made by special rays which are given out by the uranium and are able to make an impression on the plate in the same way as ordinary light. You can also test those rays in another way, by placing them on an electroscope. You know what an electroscope is. If you charge it, you can keep it charged several hours and more, unless uranium salts are placed near to it. But if this is the case the electroscope loses its charge and the gold or aluminum leaf falls gradually in a progressive way. The speed with which the leaf moves may be used as a measure of the intensity of the rays; the greater the speed, the greater the intensity.

I spent some time in studying the way of making good measurements of the uranium rays, and then I wanted to know if there were other elements, giving out rays of the same kind. So I took up a work about all known elements and their compounds and found that uranium compounds are active and also all thorium compounds, but other elements were not found active, nor were their compounds. As for the uranium and thorium compounds, I found that they were active in proportion to their uranium or thorium content. The more uranium or thorium, the greater the activity, the activity being an atomic property of the elements, uranium and thorium.

Then I took up measurements of minerals and I found that several of those which contain uranium or thorium or both were active. But then the activity was not what I would expect; it was greater than for uranium or thorium compounds, like the oxides which are almost entirely composed of these elements. Then I thought that there should be in the minerals some unknown element having a much greater radioactivity than uranium or thorium. And I wanted to find and to separate that element, and I settled to that work with Professor Curie. We thought it would be done in several weeks or months, but it was not so. It took many years of hard work to finish that task. There was not *one* element; there were several of them. But the most important is radium, which could be separated in a pure state.

All the tests for the separation were done by the method of electrical measurements with some kind of electroscope. We just had to make chemical separations and to examine all products obtained, with respect to their activity. The product which retained the radioactivity was considered as that one which had kept the new element; and, and as the radioactivity was more strong in some products, we knew that we had succeeded in concentrating the new element. The radioactivity was used in the same was as a spectroscopical test.

The difficulty was that there is not much radium in a mineral; this we did not know at the beginning. But we now know that there is not even one part of radium in a million parts of good ore. And, too, to get a small quantity of pure radium salt, one is obliged to work up a huge quantity of ore. And that was very hard in a laboratory.

We had not even a good laboratory at that time. We worked in a hangar where there were no improvements, no good chemical arrangements. We had no help, no money. And because of that, the work could not go on as it would have done under better conditions. I did myself the numerous crystallizations which were wanted to get the radium salt separated from the barium salt, with which it is obtained, out of the ore. And in 1902 I finally succeeded in getting pure radium chloride and determining the atomic weight of the new element, radium, which is 226, while that of barium is only 137.

Later I could also separate the metal radium, but that was a very difficult work; and, as it is not necessary for the use of radium to have it in this state, it is not generally prepared that way.

Now, the special interest of radium is in the intensity of its rays, which is several million times greater than the uranium rays. And the effects of the rays make the radium so important. If we take a practical point of view, then the most important property of the rays is the production of physiological effects on the cells of the human organism. These effects may be used for the cure of several diseases. Good results have been obtained in many cases. What is considered particularly important is the treatment of cancer. The medical utilization of radium makes it necessary to get that element in sufficient quantities. And so a factory of radium was started, to begin with, in France, and later in America, where a big quantity of ore named carnotite is available. America does not produce many grams of radium every year but the price is still very high because the quantity of radium contained in the ore is so small. The radium is more than a hundred thousand times dearer than gold.

But we must not forget that when radium was discovered no one knew that it would prove useful in hospitals. The work was one of pure science. And this is a proof that scientific work must not be considered from the point of view of the direct usefulness of it. It must be done for itself, for the beauty of science, and then there is always the chance that a scientific discovery may become, like the radium, a benefit for humanity.

But science is not rich; it does not dispose of important means; it does not generally meet recognition before the material usefulness of it has been proved. The factories produce many grams of radium every year, but the laboratories have very small quantities. It is the same for my laboratory, and I am very grateful to the American women who wish me to have more of radium, and give me the opportunity of doing more work with it.

The scientific history of radium is beautiful. The properties of the rays have been studied very closely. We know that particles are expelled from radium with a very great velocity, near to that of light. We know that the atoms of radium are destroyed by expulsion of these particles, some of which are atoms of helium. And in that way it has been proved that the radioactive elements are constantly disintegrating, and that they produce, at the end, ordinary elements, principally helium and lead. That is, as you see, a theory of transformation of atoms, which are not stable, as was believed before, but may undergo spontaneous changes.

Radium is not alone in having these properties. Many having other radioelements are known already: the polonium, the mesothorium, the radiothorium, the actinium. We know also radioactive gases, named emanations. There is a great variety of substances and effects in radioactivity. There is always a vast field left to experimentation and I hope that we may have some beautiful progress in the following years. It is my earnest desire that some of you should carry on this scientific work, and keep for your ambition the determination to make a permanent contribution to science.

A lecture given by Mme. Curie at Vassar College in the 1920s in appreciation of the gift from American women of a gram of radium.

Questions and Topics for Discussion and Writing

1. Why was Marie Curie so interested in radioactivity that she would devote so many years of strenuous labour to isolating and measuring radium?

2. Describe the "beauty of science" that Curie refers to. What are some of the "beautiful" aspects of her work, as she might have seen them?

3. Curie died of leukemia, doubtless caused by overexposure to radioactive substances. Before her work, the dangers of exposure to radiation were not known. Comment on the risks of such scientific investigation to the investigator. Is the knowledge gained (and the "beauty" beheld) worth the danger encountered?

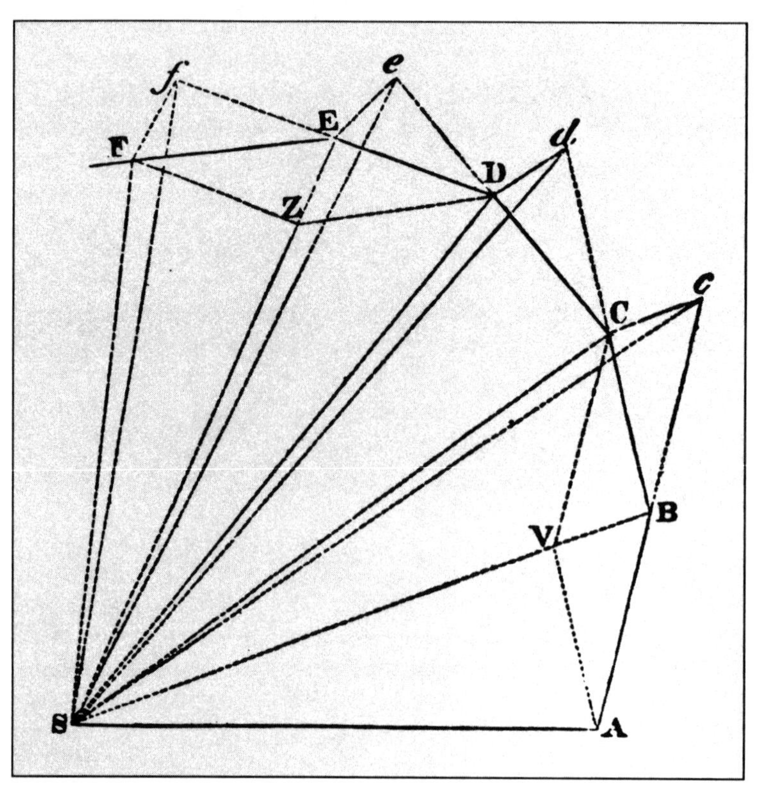

Newton's diagram of the effect of a centripetal force upon a uniformly moving body.

Issue

The Development of the Physical Sciences

The great corpus of science can be examined more closely by focusing on only a part of science. All of what is science can be divided into more manageable parts using any of a number of criteria—method of investigation, or logical structure of theories, for instance—but the most usual division is by the subject matter studied. A common hierarchy, which will also be followed here, is to divide all science into two groups: the natural sciences and the social sciences. The natural sciences are those which study the phenomena of nature, while the social sciences study the organizations and structures of human culture.

While there is much overlap of methods of study and subject matter studied, it is also true that there are vast differences between the natural sciences and the social sciences, differences due to the fact that in the social sciences human interactions present a level of unpredictability that must be allowed for and accounted for throughout all social theories. It is therefore a common practice to consider social science as something very different from the natural sciences, and to use the term *science* to refer to the natural sciences alone. That convention will also be followed here, and this book will concern itself almost exclusively with natural science and its interaction with society.

The natural sciences divide most simply and obviously into two great sections: the physical sciences and the life sciences. The life sciences are concerned with the phenomena of nature that are in some sense alive while the physical sciences are concerned with everything else. The physical sciences include astronomy, physics, chemistry, and geology, among others, and, for convenience, we shall here also consider mathematics (though many people consider mathematics not to be a science at all).

The history of the physical sciences shows a definite sequence of development of the different component sciences, a sequence that says a great deal about the structure of these sciences. Physical sciences develop rapidly and become fruitful areas of inquiry when reliable physical laws can be stated and tested, and used as the basis for discovering other laws. Most often, this has required a precise mathematical formulation. Therefore what is required is the discovery of relationships that are simple enough to be stated in mathematical terms, and mathematics that is sophisticated enough to express those relationships.

The order in which the physical sciences "came of age" is: mathematics, astronomy, physics, chemistry, and geology. Mathematics was made rigorous and thereby reliable and fruitful in antiquity, notably by the Greeks. Astronomy was a precise observational science in antiquity, but moved forward rapidly after mathematically precise and simple descriptions of the planetary motions were found in the seventeenth century. Physics was a subject of speculation, but not of measurement nor reliable mathematical formulation until the seventeenth century. Chemistry followed in the eighteenth century; geology in the nineteenth.

The articles in this section focus primarily on developments in physics, with some attention to astronomy and a mention of mathematics. Also the period of greatest attention is the sixteenth and seventeenth centuries—the period known as the Scientific Revolution. Byron Wall's essay, "What the Copernican Revolution is All About," discusses the profound shift in thinking brought about by the theory of Copernicus that it was the earth that travelled around the sun, and not vice-versa. Wall's "Proposition I.47 of Euclid: The Pythagorean Theorem" illustrates the rigor and beauty of Greek mathematics which formed the backbone of scientific theories in all the physical sciences. Thomas H. Leith sketches the life and personality of the recluse who set the Scientific Revolution in motion in "Copernicus: The Man."

Quite the opposite kind of personality was the argumentative and outspoken Galileo, who did far more to push the Copernican Revolution than did Copernicus himself. Galileo's "Starry Messenger" is one of Galileo's monographs that supported the Copernican view with observations of heavenly phenomena seen for the first time with the aid of the newly invented telescope. Pushing Copernicanism was not just a matter of arguing a point of view for the sake of knowledge alone. Galileo's advocacy flew in the face of the established dogma of the Catholic Church and brought him the wrath of the Inquisition. "The Sentence of Galileo by the Inquisition" is the decree of the Inquisition which placed Galileo under house arrest for the rest of his life, giving the reasons for its actions.

The Scientific Revolution began with the Copernican Revolution and ended with the birth or rebirth of all the major modern physical sciences. The person who drew together the various theories and discoveries of the Scientific Revolution into one great synthesis of physical theory was Isaac Newton. Thomas H. Leith's "Newton: The Man" is a brief sketch of Newton's life and very peculiar personality. Newton's synthesis survived as the foundation of physical theory until it was finally supplanted by developments in physics in the early twentieth century. The most notable physicist of this period was Albert Einstein, whose theory of relativity made physics more exact in the realm of very large distances and very fast speeds. James A. Newman's "Einstein" is an appreciation of Einstein and his work, but also an insightful essay on the workings of the scientific mind.

What the Copernican Revolution is All About

Byron E. Wall

The Copernican "Revolution," was a revolution in two senses. By attributing the "revolution" to Copernicus, reference is made to Copernicus' chief piece of writing, his book *On the Revolutions of the Heavenly Spheres*. Hence, the "revolution" is that of the earth's annual motion and those of other planets around the sun, or alternately, of the daily rotation of the earth on its own axis.

On the other hand, the more political sense of "revolution" is also intended, because the Copernican Revolution represented a turning of mankind's allegiances away from one system to another. In this case the systems were not of government, but of beliefs and modes of thought. Perhaps it would be better to go with the metaphor farther and say that the Copernican Revolution *was* an overthrow of one system of government for another, where the powers of these governments rested not in brute forces but in systems of ideas that exercised controls, first, on the thoughts of men, and thereby, indirectly, on their actions.

Ancient Astronomy

Astronomy is a very ancient science. By the standards we now use to determine what is or is not a science, astronomy was the first subject to become a science. Astronomy today is an inordinately complex field, involving assimilating data from huge and expensive optical and radio telescopes, subjecting these data to intricate mathematical analysis, interpreting the results in terms of both the modern physical theories that apply to vast distances and speeds and the theories applicable to sub-atomic particles. From this research comes information about the size and nature of the universe, its history, and perhaps its future, and in a more immediately practical form, information that will assist the development of aerospace technology. Ancient astronomy was of course very different in its specific manifestations, but in a general way its aims and even its methods have much in common with astronomy today.

Using a very inclusive definition of what constitutes astronomy, it is obvious that all early cultures must have had some kind of astronomical theory, even if only the bare observation that the sun rises and sets every day. But there is evidence that virtually all ancient civilizations went a lot farther than that.

The ancient Mayans in Central America had an advanced mathematics, an essential prerequisite for the development of astronomical theory, but their astronomy confined itself mainly to the regulation of the calendar. It's not that calendars are not important—they predict the seasons and hence the proper times for sowing and harvesting crops, and they determine the proper dates and times for religious rituals. It was indeed concern with the calendar that gave great impetus to European astronomy in the time of Copernicus. What the Mayans didn't do was apply their sophisticated mathematics to an analysis of the complex motions of the heavenly bodies.

In European astronomy, calculations of this sort were the central concerns of the subject. The accuracy of some of the calculations of planetary positions made in ancient Greece, using cumbersome mathematics and naked-eye observations, is staggering. Careful observation and painstaking calculation makes the ancient astronomy of the Mediterranean civilizations look very modern. One important difference between ancient and modern, though, was the choice of heavenly bodies studied.

Today's astronomers are primarily concerned with very distant stars. The ancients treated these as a more or less fixed background against which the objects of their attention moved.

The general idea was that all the stars, or what *we* would call stars, were equidistant from the earth, and were fixed, as it were, on a huge sphere which turned around on its own axis once every twenty-four hours, carrying everything with it. The ancients were not able to detect any movement of these stars with respect to each other, only the daily rotation of the whole troupe. What fascinated them were seven exceptions to this principle. They called them "wandering stars". We still call them by the Greek word for "wanderer"—planet. There were seven planets in ancient astronomy—Mercury, Venus, Mars, Jupiter, Saturn, and also the Sun and the Moon. They all had these characteristics in common: they made a cycle through the sky every twenty-four hours, just like the fixed stars, but they did not exactly keep pace with them. They had, it seemed, a motion of their own, in addition to partaking of the motion of the sphere of the fixed stars. Moreover, the paths of their motions relative to their fixed counterparts were all in a narrow band that cut diagonally across the large stellar sphere. That narrow band was where all the action took place. The stars that were fixed in it became the landmarks by which the wanderings of the planets could be detected. Since the sun took one year to complete its journey around the band, the band could be divided into twelve sections, so that the sun spent a month, more or less, in each one. In each of the twelve sections, the brightest stars formed a recognizable pattern, or constellation, each of which was given the name of the animal figure, or in Greek, *zoidion*, that the constellation was purported to resemble.

Thus, what *ancient* astronomy, at any rate, was all about, was the movements of seven spots of light through the signs of the *zodiac*. Of what value was this arcane study other than the opportunity it provided for mental gymnastics? If we confine the subject to the rotation of the sphere of the fixed stars, including the very slight deviation or wobble of the sphere in its daily motion, which was noticed by the ancients and carefully recorded, plus the study of the motions of the sun and moon as they move against the fixed stars, then we have an adequate basis for regulating the calendar, making agricultural decisions that depend on the seasons, for navigating land and sea journeys by the stars, and even for accurately predicting eclipses. But the most complicated studies were those devoted to accounting for the motions of Mercury, Venus, Mars, Jupiter and Saturn. It is not easy to find a utilitarian function for the results of the study of these planets. At least, it is not easy if we insist on evaluating their work from our own worldview. The fact that the objects of their study were planetary motions through the signs of the zodiac alerts us to the significance of their work for astrological usages. If the positions of the planets among the signs of the zodiac were held to influence the destiny of individuals on earth, then the precise determination of those positions in the past and future could be considered extremely important. Indeed, the terms astrologer and astronomer were virtually interchangeable throughout the ancient and medieval world. Casting horoscopes provided a considerable source of income for astronomers in pre-modern times.

Helpful as the horoscope industry may have been in providing livelihoods for astronomers, the concerns of astrology were not major factors in the Copernican revolution. The justification for ancient planetary astronomy that *was* crucial later in the Copernican era had to do with the overall cosmology of Greek philosophy.

Aristotle's Cosmology

The dominant feature of Greek philosophy was its tidiness—everything in its place in a coherent order. This was nowhere more evident than in cosmology. The cosmological theory that had the greatest acceptance was Aristotle's. According to Aristotle, the world was divided into two neat compartments. The inner compartment was the world of generation and corruption. It was the world we lived in, where things grew and died and moved around in a seemingly complex disorder. This apparent disorder was put in philosophical order by analyzing material things into combinations of earth, air, fire and water, and attributing natural tendencies to each of these elements. Things composed of earth and water moved downward, that is, inward toward the centre of the world, because of their inherent heaviness. Fire and air moved up-

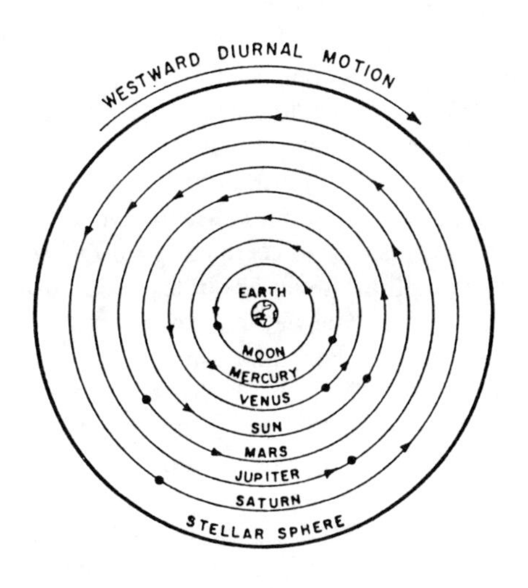

FIGURE 1.1: Aristotle's universe of concentric spheres.

ward, that is, away from the centre because of their inherent lightness. Forces which interfered with these natural motions were attributed to external, violent causes. A good example is that of a stone thrown upward: its original upward motion is due to the force of the agent that threw it. When the force is expended, the stone resumes the motion inherent in its heavy nature and falls to the ground. By such arguments was every physical event in the corruptible world accounted for. The corruptible world included the earth itself and everything between it and the moon, which was believed to be the closest of the planets.

But from the moon outward to the fixed stars, everything was different. Out there things did not grow and die, nor were their motions random. Things neither fell toward the earth nor floated farther and farther away from it. They just went round and round, day after day and year after year. By the way, meteors, shooting stars, comets and other such items involved in unique occurrences were believed to be all below the moon. The physics that explained motions in the world around us just would not do for motions in the heavens. Hence for that outer world, a completely different physics was expounded.

Up beyond the moon, there was no earth, air, fire or water, but a fifth element, or quintessence,—a crystalline substance that filled all space. To this fifth element, only one kind of motion was natural: circular. The motions were circular for two reasons: (1) It was what they seemed to be, in general. The stars went round in circles, night after night; the sun came up in the east, set in the west, then came up in the east the next day, and so on. (2) Circular motions have eternity built into their definitions. Motions in straight lines require beginnings and endings, generations and corruptions. Circular motions require no starting and stopping. So we get a nice tidy system of celestial mechanics if everything up there goes obediently around in circles.

The Problem of the Planets

But here we come smack up against the problem of the planets. These mysterious wanderers threaten to upset the whole system. Not only do they not go nicely around in simple circles in step with the other stars, not only do they run a course in approximately the opposite direction, they actually seem to start and stop, then turn around and go in the other direction for brief periods of time. Sometimes they actually seem to do loops in the sky. The only saving grace is that they seem to perform these antics over and over again, so that there must be some regular order to their motions.

The task for astronomers, then, was to find out how to account for the apparent motions in terms of circular motions, perhaps as the result of several circular motions compounded together. This is the task that astronomers were charged with by philosophers. Plato referred to it as "saving the phenomena"; it might have been more appropriately called "saving the philosophy."

Aristotle proposed his own solution to the problem of the planets, adapted from an idea of another Greek thinker. In Aristotle's version, each planet was on a different spherical shell, all shells being concentric with the earth. Between the shells on which a planet was stuck were intermediate shells. Except for the planets, which were each on different shells, and the fixed stars, which were all on the nearly outermost shell, the rest of the substance of the shells was invisible.

The outermost shell was the prime mover, which somehow communicated its motion to the next shell, but the next shell rotated on its own in a slightly different direction, so that anything visible

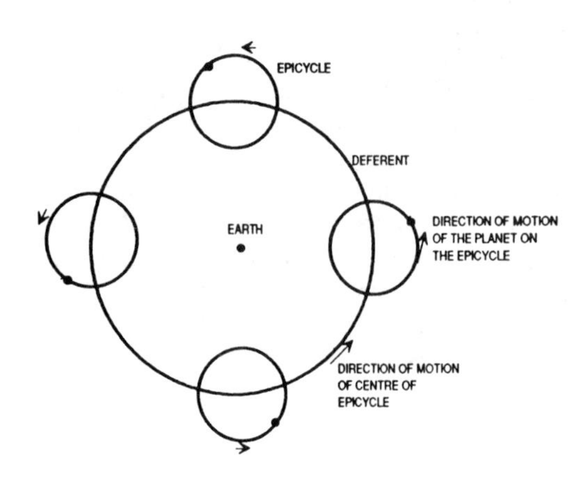

FIGURE 1.2: A planet travelling on an epicycle, which is fixed on a rotating deferent.

on it would move with the combined motions of the outer shell in one direction and its own shell in another direction. That shell in turn carried with it the next shell, which also rotated in a different direction, and so on. Thus visible points on the surfaces of these shells, such as planets, would appear to be moving quite irregularly, but actually their motions were merely the combined effects of numerous circular motions from the rotations of other shells.

The idea was ingenious in principle, but unfortunately it could never be made to work very well. It just did not seem possible to think up an arrangement of shells that could move in such a way as to specify the planetary motions that had been observed and recorded.

The System of Ptolemy

By the end of the second century A.D., another conception for accounting for the planetary motions had been worked out. This was the system perfected by Claudius Ptolemy, a Greek astronomer whose work we have come to know by its Arabic title, *Almagest*. It was the general system of Ptolemy that was the accepted basis for astronomy in the time of Copernicus, and it was in particular Ptolemy's work, *Almagest*, that Copernicus directed his work against.

Ptolemy's system accepted the basic Aristotelian conception for the "inner" world between the moon and the earth, and also the general idea of the sphere of the fixed stars that rotated every twenty-four hours. He was also able to account for the motions of the planets with combinations of circular motions. But instead of postulating a set of concentric spherical shells, like onion skins revolving every which way, his way of accounting for the irregular planetary motions involved circles which turned on circles. Arthur Koestler has called it the Ferris Wheel universe. For each planet, one big circle called the deferent described a path through the zodiac that represented the general direction of the planet's motion. Somewhere on the edge of the deferent was the centre of a second, generally smaller circle called the epicycle. At the edge of the epicycle was the actual planet. Both the deferent and the epicycle turned, at rates specified by Ptolemy for each planet, which resulted in the planet moving in what appeared from the vantage point of the earth to be the irregular combinations of forward, backward, starting and stopping, and looping motions that had so troubled Greek philosophy. But of course in Ptolemy's system these irregular appearances were actually to be just the visual result of combined smooth, circular, and of course, eternal, motions.

The great advantage of Ptolemy's system over Aristotle's was that it worked—it could indeed account for the planetary motions within a reasonable margin of error. It was pretty complicated though. The concept of deferents and epicycles had been around for a long time before Ptolemy had worked it into a reliable system. In getting the bugs out, he had to introduce a number of inelegant details. For example, the centre of the deferents were not necessarily the centre of the earth; they could be taken to be eccentric to the earth, which would explain why a planet would seem to complete one-half of its course in a shorter time than it would take to travel through the other half. For similar reasons, Ptolemy did not require that the epicycles turn at uniform speeds around their respective centres, as long as they turned uniformly around some other chosen point within their circumferences.

But, after all the necessary adjustments, Ptolemy did come up with a way of accounting for the planetary positions that had been observed and

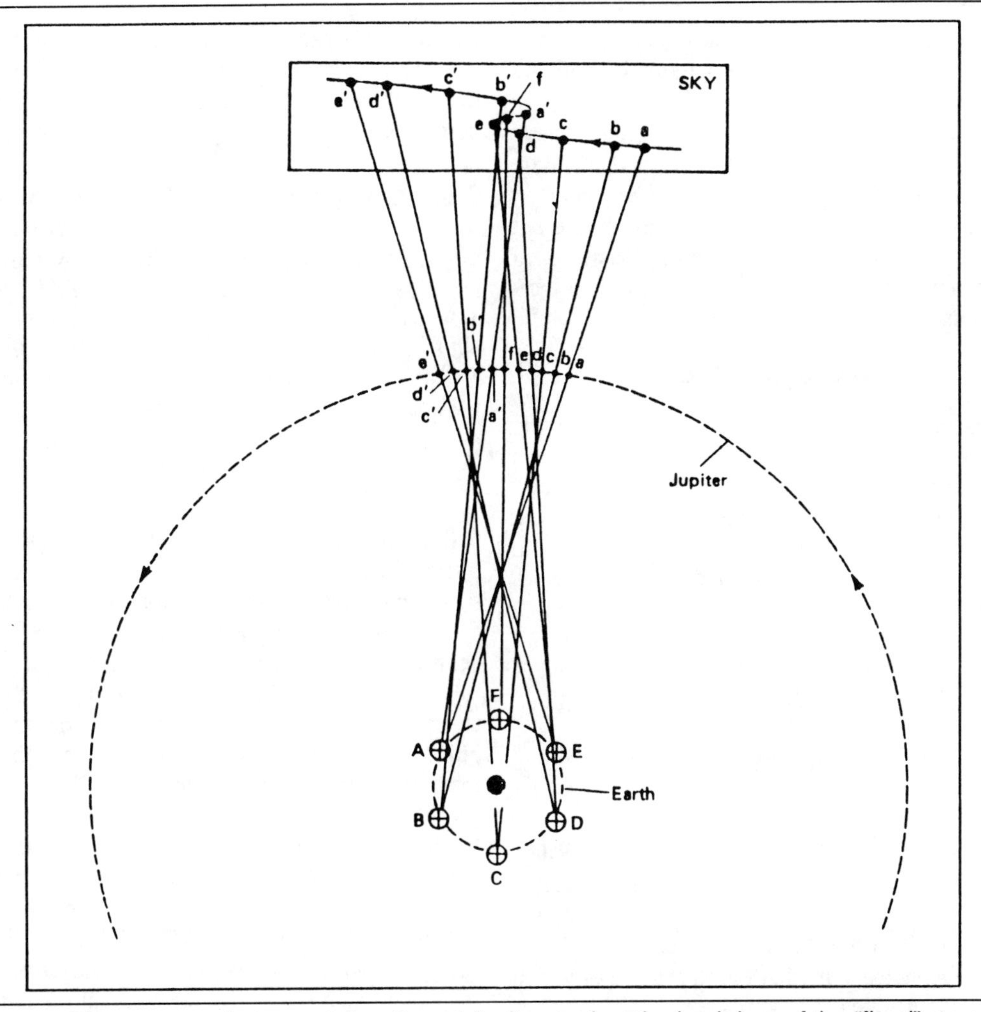

FIGURE 1.3: The apparent "retrograde" motion of Jupiter against the backdrop of the "fixed" stars can be seen to be an illusion caused by the combinations of the motions of the Earth and Jupiter around the Sun.

predicting future positions of the planets as they would appear against the stars in the zodiac. The philosophical requirements of the perfect heavens had been fulfilled. The planets were seen to move in never-ending circular pathways at what could be at least construed as uniform, that is, unchanging, speeds. It was regular enough that it was possible to construct a mechanical clock-work system that would show the positions of the planets as the gears turned round. In the Middle Ages, just such a clock was built along Ptolemaic principles.

The real fault of Ptolemy's system was that despite meeting the philosophical requirements that had been specified, the system just didn't make philosophical sense. It's all fine to specify precise combinations of circular paths and speeds for the planets and to show that they don't really start and stop, but even in Ptolemy's plan, the planets are supposed to be really doing loops in the sky as they pass through the zodiac. Why should they be doing that?

In Aristotle's onion-skin system, the planets were carried around on solid pieces of stuff. The way they were held up and moved around had some logic to it, even if the motions themselves didn't. Ptolemy had the motions, but no explanation for them—no explanation for what was moving the planets around and no explanation for why they should be moving in this peculiar fashion.

The problem of what was moving the planets around was a subject of much study in the Middle Ages and Renaissance. It isn't just one "clock" on which all the planets are shown, but seven different clocks, one for each planet. The motions of each planet are conceived differently. There is no principle to unify them.

The Copernican Viewpoint

It is the work of Copernicus that provides the unifying principle. By taking the sun, or rather a point near the sun, as the main centre of the motion of the planets, he could dispense with the epicycles, for their main function. By taking the sun as fixed, and making the earth into a wanderer like the other planets, the epicycles that Ptolemy needed can be seen to be just an artifact of our perception due to the combination of the motions of ourselves and of the planets all around the sun. The planets don't do loops and other dances in the sky, they just appear to, because as we watch their motions around the sun, we see them from different angles due to our own motion. The vast sphere of the fixed stars doesn't spin around at enormous velocity every day, instead we revolve on our own earthly axis, and that makes the stars appear to move. Moreover, all that is special about the stars in the zodiac is that they happen to be the stars that fall in the line of vision between ourselves and the planets.

But now we come up against the more political sense of revolution in the Copernican revolution—the overthrow of a complete system of ideas. In the Aristotelian worldview, even as amended by Ptolemy, we lived in the centre of generation and corruption, of living and dying, stopping and going, in short, in an imperfect, temporal world. Out there, beyond the moon, everything is perfect and eternal and changeless. But if *we* are also eternally on a celestial body, does that mean that we are also eternal, perfect, incorruptible, and, for that matter, weightless? No, that can't be, so it must then be that there is no sharp distinction between the perfect and imperfect worlds. Maybe it's all like it is here on earth. But then why do the planets move eternally around the sun? Why is it that when we let go of something heavy it doesn't fly away to the sun if the sun is the centre? Why don't birds and clouds disappear toward the west if we are spinning around toward the east? If the heavens that we see aren't perfect, where is perfection to be found? Where is heaven?

These questions put to the Aristotelian worldview were not just idle analyses of the thought of some ancient philosopher. Aristotle's system was taken over by the Church during the Middle Ages and made into its own. Scripture had been interpreted in an Aristotelian framework. God's world was to be understood with the help of the heathen philosopher. The heavens were supposed to be telling the glory of God, and reminding us of our imperfections down here on earth. Now what were we supposed to think?

Copernicus was not unaware of some of these difficulties. He offered some possible explanations for the more paradoxical questions about terrestrial physics on a moving earth which was not at the centre of the world, and gave a rather poetic justification for the sun being in the centre. But thanks to an overzealous colleague, the Lutheran theologian Osiander, these more shattering questions and the answers they lead to could be quietly put off for a while. When Copernicus' book appeared, Osiander had anonymously added a preface saying that the sun-centered system given in the book was to be viewed as merely a convenience for purposes of astronomical *calculations*, and need not be taken literally. And that is precisely what was done with the Copernican system for the next half-century or more, with some notable exceptions including Giordano Bruno, who dared to take the system literally and draw too many embarrassing implications from it, and was burnt at the stake in consequence.

Sooner or later the issue had to be faced; it had to be laid at the feet of the general educated public and be discussed and analyzed in detail. Until then *astronomers* could use the Copernican system in their calculations and ignore the consequences of taking it literally. Non-astronomers who could not appreciate the technical reasons favouring the Copernican system could easily dismiss the notion of the earth as a moving planet as utter nonsense, and did just that.

Because the Aristotelian-Ptolemaic conception of the world was so entrenched, it was accepted as just an elaboration of common sense—which in a way it was, of course. It was therefore accepted not just as a good theory, but as the simple truth. So self-evident did the immobility of the earth seem, that for the general public to abandon that view, a complete revolution of their normal thinking would have to take place. In the sense of revolution here intended, the monumental work of Copernicus was hardly a suitable vehicle for bringing it about, despite the appearance of the word "revolution" in the title of Copernicus' book.

On the Revolutions is, except to those with a taste for tedious mathematical calculation, a very dull, stupefying piece of writing with very little to capture the imagination, much less launch a revolution in human thought by the boldness of its polemic. This is a point that has been repeatedly emphasized to me by my students whenever I have assigned them to read sections of it. Even granting the adventurous climate of the Renaissance and the newfound respect for ideas that permeated the revival of learning, it is hard to see how the Copernican view could begin to get serious consideration outside of mathematically astute circles—unless it was pressed very vigorously by a very strong personality.

Galileo, the Revolutionary Leader

As it turns out, that is exactly what happened in Italy in the first decades of the seventeenth century. The strong personality was Galileo. Galileo was astute, bold, fearless—one might even say reckless. He used all the tricks known to debating societies, encyclopedia salesmen and movie stars to sell his product, the Copernican system, which he happened to believe in. And by and large, he succeeded.

He took apart the arguments of his opponents, the Aristotelian scholars, and made laughing stock of them. He produced new and totally unsuspected evidence that defied assimilation into the Aristotelian conception, but which fit nicely into *his* view. He managed to get the book in which he set out his strongest arguments banned, which of course only increased the demand for it. He even employed the technique for becoming famous, which we in the post-Watergate era now know to be flawless, of getting himself convicted for a major crime.

The new evidence came from the telescope, then recently invented and much improved by Galileo. With it he discovered many new facts about the heavens which undermined the Aristotelian conception. The moon, for example, he could see to be rocky and uneven. It had mountains, which Galileo calculated to be as large as those on earth; it had craters; it had dark spots which Galileo mistakenly thought to be seas. It was therefore much like the earth in its irregular configuration. In other words, it was not smooth and perfect as fit a heavenly body according to the Aristotelians.

He also noticed that besides the illumination of the moon by the sun, the moon had a secondary light that must have come via reflection of the sun's light off the earth. And of course, the earth has a rough and irregular surface. This answered the objection that if the heavenly bodies were not smooth and polished, they could not reflect light.

The visible phases of the moon were not important for the Copernican system, as the moon was held to go around the earth in any case. But phases of the *planets* would be another matter. With his telescope, Galileo was able to see that Venus had phases like the moon's. These could *only* be explained if Venus travelled around the sun. Yet in the Ptolemaic system, Venus was placed between the sun and the earth at all times.

The placement of the moon in the Copernican system was an anomaly. If all the planets were supposed to go around the sun, why did the moon go around the earth instead? Galileo was able to forestall this objection by finding through his telescope that the moon was not an exception. Jupiter also had *moons* that circled it—four of them that he could see.

Galileo published these and other findings in a short book called *The Starry Messenger*. In Galileo's book the observations are carefully reported, and the reader is told how to construct a telescope so he can go and see for himself. The book caused a stir and prompted some "refutations" to the effect that the observations were impossible and mistaken or were caused by distortions in the telescope itself. A breach had been made, but there were still plenty of defences available.

Galileo answered his opponents, published other findings, invited people to use his telescope under his guidance, and eventually began to address himself to issues that were not strictly astronomical, but were concerned with implications in other areas, for example whether there was a necessary connection between Aristotelian cosmology and Christian theology. This was dangerous ground. Galileo gained the reputation of a trouble-maker. He was warned by the Vatican. For a time he kept quiet, but during this time he worked out his next arguments and how they would be presented.

Eventually he published the book which got him in deep trouble, *The Dialogue Concerning the Two Chief World Systems*, by which is meant the Copernican versus the Aristotelian-Ptolemaic systems. The *Dialogue* is a lengthy book, written in the style of a Plato dialogue, with three characters: a mathematician-scientist who espouses the Copernican system, an Aristotelian scholar who defends the status quo, and a neutral interlocutor who is open to being convinced one way or the other. Galileo wrote the *Dialogue* in Italian, not in Latin, which meant that he was deliberately playing to the galleries, that is, to the ordinary educated citizen rather than to the scholars, who he viewed as too entrenched in their fixed ideas.

As one would expect, the mathematician-scientist has no trouble destroying all the objections of the Aristotelian while making a flawless case for Copernicus. The Aristotelian, who is called Simplicio, is revealed as a true simpleton who again and again argues himself into absurdities, fails to see the point and blithely continues to maintain his original position. The neutral observer slowly is won over to the Copernican view after having made numerous reasonable objections which he eventually sees are satisfactorily answered within the Copernican system.

In 1632 the *Dialogue* was published. Five months later it was banned. Galileo was called before the Inquisition, tried and pronounced to be vehemently suspected of heresy, ordered to recant his false opinions, and was placed under house arrest for the remainder of his life.

The Basis of Knowledge: Experience versus Authority

Galileo lost his battle for Copernicus, but of course the war was eventually won, and Galileo's fate at the hands of the Inquisition *promoted,* rather than retarded his cause. It did so because it pinpointed the crucial issue that makes the word revolution in the Copernican Revolution refer to more than planetary motions. That issue had to do with whether knowledge, at least knowledge about the external world, rests on a basis of sense experience or on the basis of authority. In this sense, the motions of the earth and the planets merely provided the test case.

Aristotle's system *was* based on sense experience, the sense experience of ancient Greece—a point Galileo repeatedly emphasized. But what was true for Aristotle was not true for the Scholastic Aristotelians, for whom the basis of knowledge was the set of books, including Aristotle's, that established the framework into which all experience had to fit. As only a few had the requisite learning to have mastered these texts, it was their lot to explain the universe to those who did not enjoy this direct access to the truth. Those who gained more of their knowledge through their senses in the daily experience of living were placed in the position of having to disregard their experience or to interpret it according to well worked out principles understood only by the scholars. That is why it was revolutionary for Galileo to address himself to the larger audience of laymen by writing in Italian instead of Latin.

Galileo's *Dialogue*, the work that we take to be the most influential in convincing the public of the truth of the Copernican world-view, was actually not pro-Copernican so much as *anti*-anti-Copernican. There are two reasons for this. One was that Galileo had received specific instructions from the Vatican not to hold or defend the Copernican system in print. Galileo interpreted this to mean that he was *not* forbidden to attack the position of those who opposed Copernicus. As we know, the Inquisition did not agree with Galileo on this fine distinction.

The second reason provides us with a glimpse of the motive that led Galileo to go to all this trouble at all. Galileo defended the theory of the long-dead Copernicus because he believed it true. Granting that, we can take it a step further and say that Galileo concerned himself with the theory of Copernicus because, to the extent that he could show the Copernican system to be true, he could thereby show the Aristotelian theory to be false. If the Aristotelian world-view could be seen to be false, then the whole authority of the ancients collapsed.

What Galileo wished to do was break the stranglehold of Aristotelianism, and it suited his purposes to use Copernicus to do it. That *this* was his major interest becomes clear when we examine Galileo's own contributions to astronomy. With the telescope, Galileo made a number of observations that lent support to the general idea that the earth moves around a stationary sun. But Galileo did not concern himself with detailed, precise measurements of heavenly motions the way that astronomers of his own era or even in antiquity did. Similarly, Galileo made no attempt to advance Copernican theory by working out a mathematical account of his new observations. In contrast, say, to his contemporary, Kepler, who advanced Copernican astronomy by discovering that the planets moved not in circles, but ellipses, Galileo expounded a simplified version of the Copernican theory in the *Dialogue* that did *not* fit the available observational data at all. Copernicus had hoped to show that his system was more accurate than Ptolemy's. The version that Galileo presented to his readers would have been much worse. But none of that mattered if all Galileo wished to accomplish was to show that the ancient theory was built on sand.

Galileo was particularly incensed about grand, sweeping theories that were unthinkingly accepted by his contemporaries. More than anything else, he wished to show that these theories are sustained only by ignorance. In a passage in the *Dialogue* the supposedly impartial observer remarks:

> There is not a single effect in nature, even the least that exists, such that the most ingenious theorists can arrive at a complete understanding of it. This vain presumption of understanding everything can have no other basis than never understanding anything. For anyone who had experienced just once the perfect understanding of one single thing, and had truly tasted how knowledge is accomplished, would recognize that of the infinity of other truths he understands nothing.

For Galileo the way to knowledge was to build it up bit by bit from observing how nature works, not by imagining great schemes. The major thrust of Galileo's scientific work that was not concerned with the Copernican view is to show how this is accomplished. In the *Dialogue*, there is enough of an indication of Galileo's conception of the road to knowledge for the reader to get the general idea. In his later book, *Two New Sciences*, that Galileo wrote while under house arrest and then smuggled out to be published in a Protestant country, that pathway is shown much more explicitly.

The way is through controlled experiment and testing of hypotheses. It relies on the data of sense perception, but with a *refinement* that made the Copernican system an excellent illustration for Galileo. One learns ultimately through the senses, but one must not rely on *common* sense for understanding. For the principles of nature to be grasped, one must experiment and reason very carefully to avoid being misled by appearances. It is *common* sense that supports the Aristotelian world-view, and only reasoning carefully from the observations that supports Copernicus.

The Triumph of Empiricism

The trial of Galileo had its effect. Italy had been a great centre of scientific activity. After the trial, science in Italy was severely retarded, and has never quite recovered. But elsewhere in the world, the trial showed those who had nothing to fear from the Inquisition that reliance on the authority of ancient philosophy leads nowhere. The new philosophy that took its place was empiricism. Its scientific basis was experiment and reason, and its watchword was to be wary of unwarranted generalizations.

Galileo had a comrade in arms for his battle for empiricism against authority in his English contemporary, Sir Francis Bacon. They had at least this much in common, but they differed on many particulars, belief in the Copernican system being one of them. More generally, Bacon distrusted the use of mathematics and logic to advance from direct observation, fearing that it would lead to the same kind of grand systems that both he and Galileo despised. But Bacon was eloquent on the value of collecting empirical data and, by a process of induction, discerning true scientific principles. Bacon's writings became a manifesto for the kind of science that Galileo had fought for with the Copernican theory as a weapon.

The end of the Copernican revolution comes with Sir Isaac Newton in England in the late seventeenth century. It was Newton who finally made physical sense of Copernicanism by showing how the new laws of physics developed by Galileo for terrestrial phenomena account for the motions of the planets as they had been described by Kepler's improvements on Copernican theory. Newton's unified physics vindicated Copernicus' celestial revolutions and established a firm basis for the new philosophy of empirical science.

That is the end of the story of what the Copernican revolution is all about, but there is one final irony. Galileo used Copernicus as a lever to overthrow the stifling authority of Aristotle. But Newton used Galileo to establish Copernicus and to lay down a new great system that engulfed all science under its authoritative and comprehensive laws. Bacon, arguing the sterility of ancient Greek science, was equally ready to dispense with the best developed part of Greek scientific thought, mathematics. But Newton, his own countryman, completed the rout of ancient science by creating a *mathematical*, deductive basis for the new science. In all revolutions, it is easier for the reins of power to change hands than it is to change the reins of power.

Questions and Topics for Discussion and Writing

1. In what way did the Aristotelian world view make more "sense" than the Copernican system?
2. Why did Aristotle have one set of rules (one *physics*) for the world "below the moon" and a different set for the world beyond the moon?
3. In what way was the system of Ptolemy an improvement over Aristotle's, and in what way was it not? In what way was the system of Copernicus an improvement over Ptolemy's, and in what way not? In what way was the system of Copernicus *not* an improvement over Aristotle's system?
4. What is the role of Galileo's own astronomical observations in establishing the Copernican system?
5. Discuss the last sentence of the article with respect to the Copernican revolution. What are the "reins of power" that have changed hands? How have the "reins" changed and how have they not changed?

Byron E. Wall

Proposition I.47 of Euclid: The Pythagorean Theorem

Byron E. Wall

The crowning achievement of ancient Greek science was the development of rigorous mathematics. The mathematics known to ancient Greeks is astonishing in its complexity, subtlety and, above all, its logical rigor. The form in which most of what we know as Greek mathematics came down to us is the thirteen books of Euclid's *Elements*, written about 300 B.C. Euclid attempted to encompass most of the mathematics known in his time in a single multi-volume treatise which began from simple axioms and developed each successive mathematical theorem, or "Proposition," from the axioms and previous propositions, using only logical inference. The approach had its limitations, not the least of which arose from Euclid's characterizing all mathematics, including number theory, as geometry, but overall the *Elements* was so successful that it remained the standard, indeed the only, textbook of geometry used, in various translations, for two thousand years. Many of the world's greatest scientists and mathematicians can trace their decision to pursue a career in science to their studying Euclid's *Elements* and becoming enthralled by the beauty and elegance of exacting mathematics.

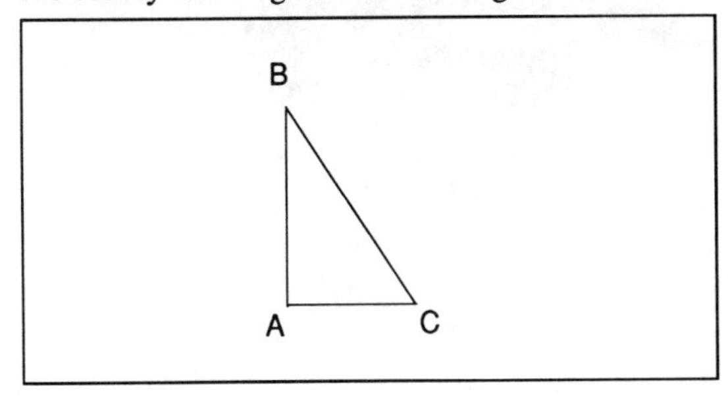

FIGURE 1.4: A right triangle.

By far, the most famous result of Greek mathematics is the theorem we ascribe to Pythagoras and which appears in Euclid as Proposition 47 in Book I. For hundreds of years this "Pythagorean" theorem was known simply as Proposition I.47. It states the familiar result, used constantly in modern science, that the area of a square constructed on the hypotenuse of a right triangle (that is, on the side opposite the right angle) is equal to the sum of the areas of squares constructed on the other two sides.

To prove this, Euclid first establishes several prior results (in Propositions I.1 to I.46), including such notions, familiar to anyone who has studied plane geometry, as (1) that triangles are congruent if one triangle has two sides and the angle between the sides equal in size to corresponding sides and

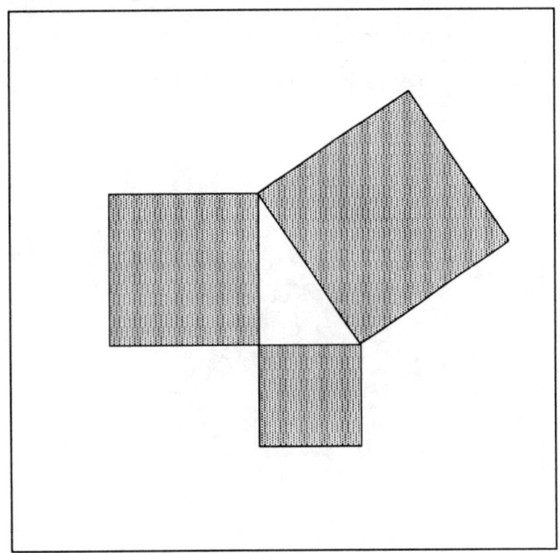

FIGURE 1.5: The squares on the sides of a right triangle.

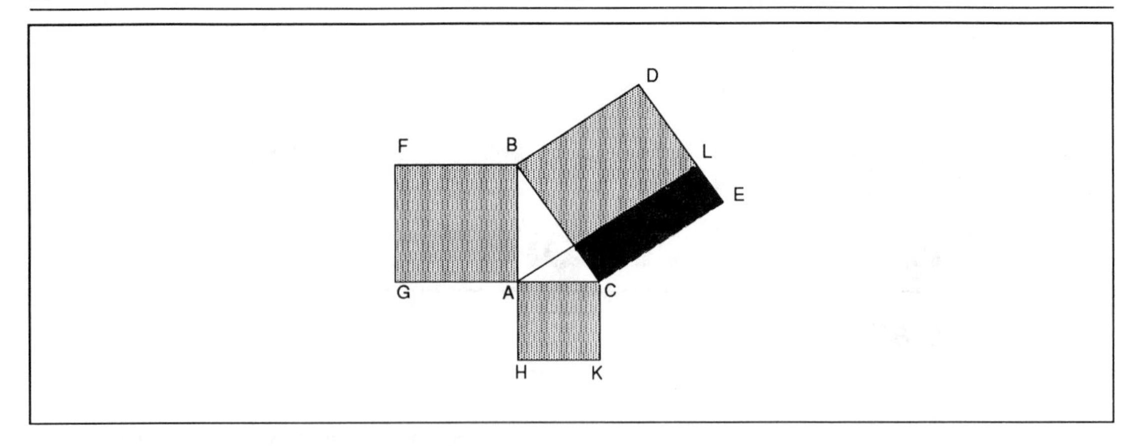

FIGURE 1.6: The square on the hypotenuse divided into two rectangles by a line from the vertex of the right angle, parallel to the sides of the square.

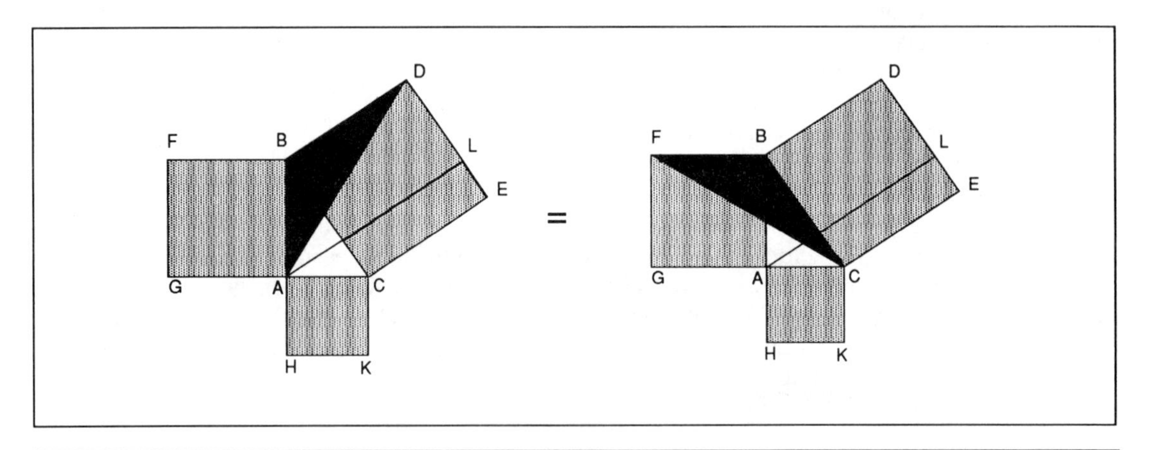

FIGURE 1.7: Triangle ABD is congruent to triangle FBC.

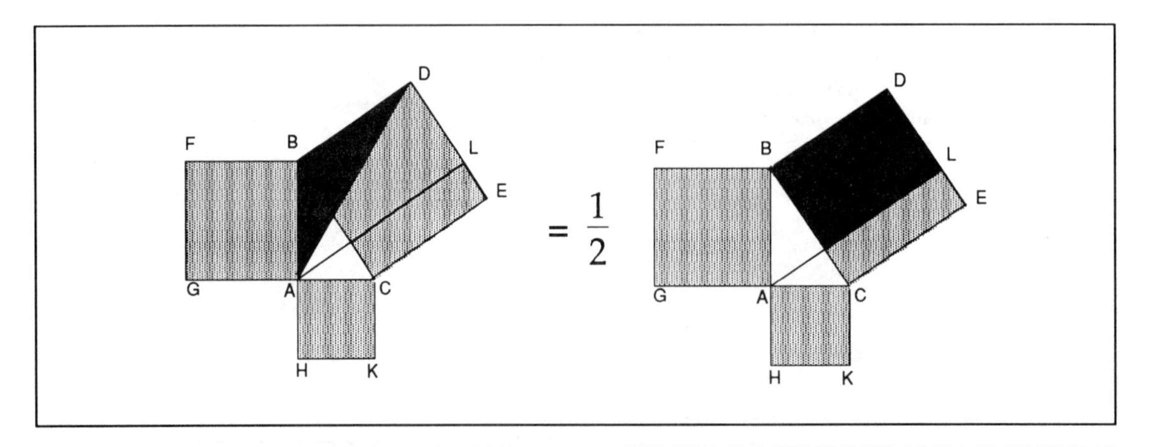

FIGURE 1.8: Triangle ABD is equal in area to ½ of rectangle BL.

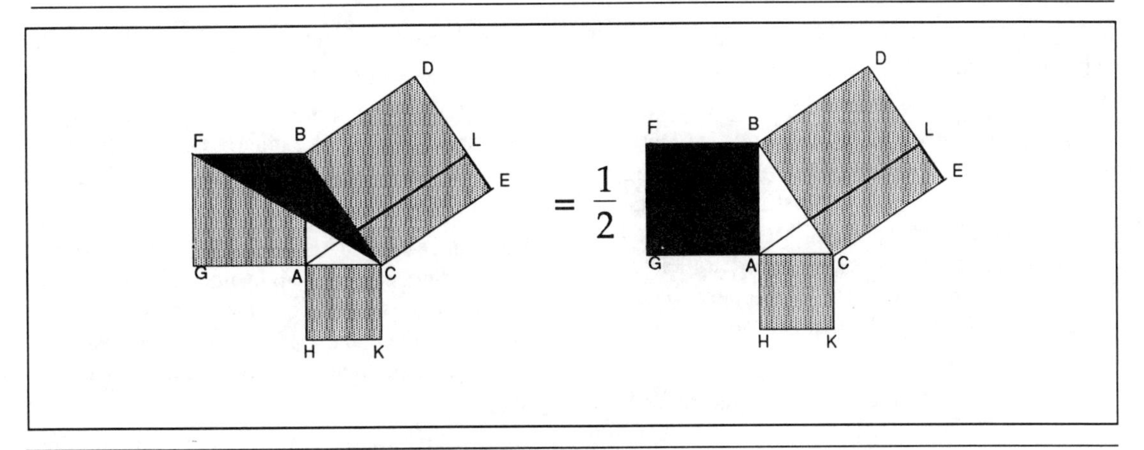

FIGURE 1.9: Triangle FBC is equal in area to ½ of square FA.

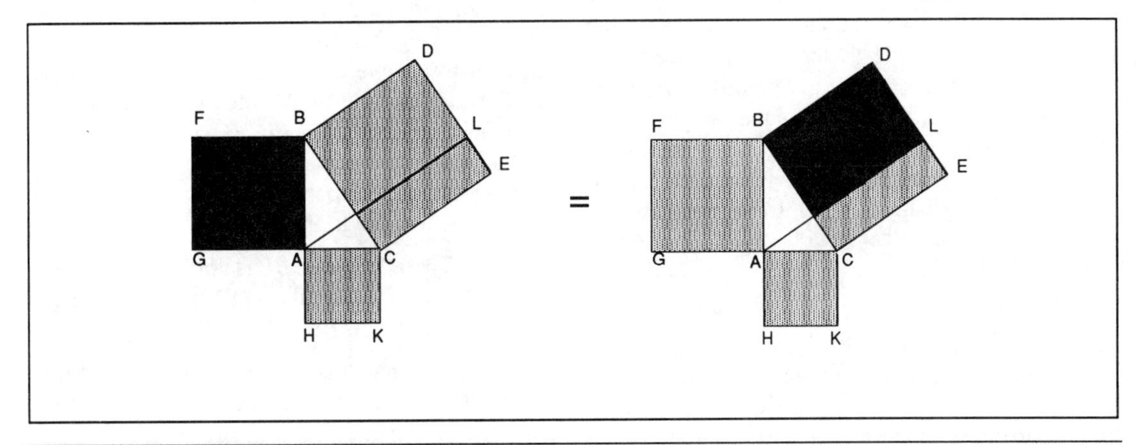

FIGURE 1.10: Square GB is equal in area to rectangle BL.

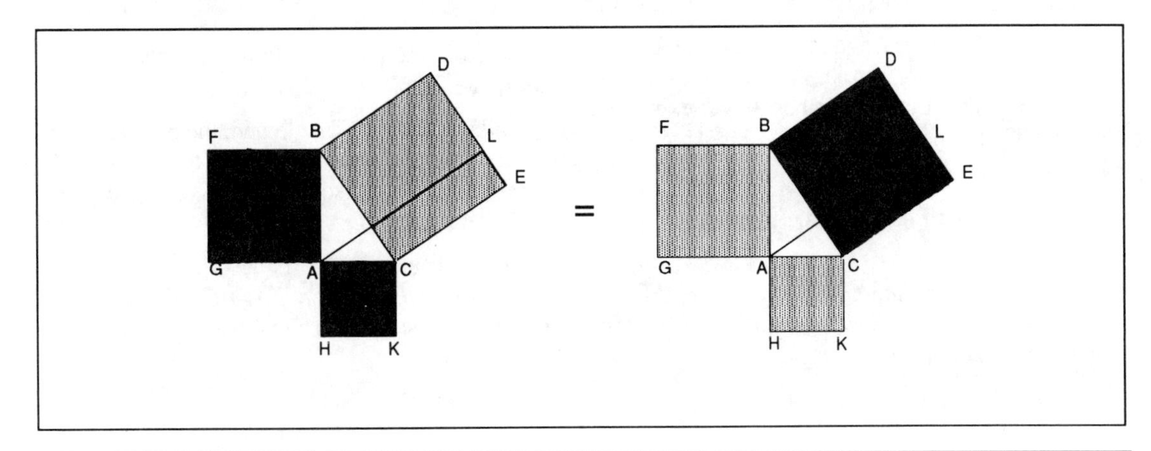

FIGURE 1.11: Squares GB and HC together are equal to square BE.

angle of the other triangle, and (2) that the area of a triangle is exactly one half that of a parallelogram with the same base and height.

His proof is very general and applies to *any* right triangle, but it is easier to see by visualizing an example. (See Figure 1.4.) First he constructs the squares on each side (Figure 1.5), then draws a line from the vertex of the right angle parallel to the sides of the square on the hypotenuse. Thus he divides the square on the hypotenuse into two rectangles. (Figures 1.6.)

The thrust of the proof is to show that each of the *other* two squares is equal in area to one of the rectangles which together make the larger square. In the most interesting and ingenious part of the proof, Euclid draws lines connecting vertices of the original right triangle to the far vertices of the squares opposite them, thus creating triangles which have bases and heights equal, respectively, to those of the rectangles which form the larger square and those of the other two squares. Euclid then shows that pairs of these (new) triangles are congruent to each other (and therefore have equal area). (Figure 1.7.)

Since squares and rectangles are also parallelograms, the area of each triangle is exactly half of the area of one of the squares or rectangles. (Figures 1.8 and 1.9) Therefore each smaller square is equal in area to one of the rectangles, (Figure 1.10) therefore the sum of the areas of the smaller squares is equal to the area of the larger square. (Figure 1.11.)

The result is magical and of immense usefulness. The proof is elegant but not at all obvious until it has been pointed out. Here is how it was expressed by Euclid, in the translation by Sir Thomas L. Heath:

PROPOSITION I.47

In right-angled triangles the square on the side subtending the right angle is equal to the squares on the sides containing the right angle.

Let *ABC* be a right-angled triangle having the angle *BAC* right; I say that the square on *BC* is equal to the squares on *BA*, *AC*.

For let there be described on *BC* the square *BDEC*, and on *BA*, *AC* the squares *GB*, *HC*; through *A* let *AL* be drawn parallel to either *BD* or *CE*, and let *AD*, *FC* be joined.

Then, since each of the angles *BAC*, *BAG* is right, it follows that with a straight line *BA*, and at the point *A* on it, the two straight lines *AC*, *AG* not lying on the same side make the adjacent angles equal to two right angles; therefore *CA* is in a straight line with *AG*.

For the same reason *BA* is also in a straight line with *AH*. And, since the angle *DBC* is equal to the angle *FBA*: for each is right: let the angle *ABC* be added to each; therefore the whole angle *DBA* is equal to the whole angle *FBC*. And, since *DB* is equal to *BC*, and *FB* to *BA*, the two sides *AB*, *BD* are equal to the two sides *FB*, *BC* respectively; and the angle *ABD* is equal to the angle *FBC*; therefore the base *AD* is equal to the base *FC*, and the triangle *ABD* is equal to the triangle *FBC*.

Now the parallelogram *BL* is double of the triangle *ABD*, for they have the same base *BD* and are in the same parallels *BD*, *AL*.

And the square *GB* is double of the triangle *FBC*, for they again have the same base *FB* and are in the same parallels *FB*, *GC*.

[But the doubles of equals are equal to one another.]

Therefore the parallelogram *BL* is also equal to the square *GB*.

Similarly, if *AE*, *BK* be joined, the parallelogram *CL* can also be proved equal to the square *HC*; therefore the whole square *BDEC* is equal to the two squares *GB*, *HC*.

And the square *BDEC* is described on *BC*, and the squares *GB*, *HC* on *BA*, *AC*. Therefore the square on the side *BC* is equal to the squares on the sides *BA*, *AC*.

Therefore etc. Q.E.D.[1]

---

[1] All Euclid's proofs end with "Therefore etc.," meaning that one can now repeat the proposition which was stated at the beginning, this time not as a proposition, but as a proved theorem. The traditional initials at the very end, "Q.E.D.," stand for "*quod erat demonstrandum,*" "that which was to be proved."

Copernicus: The Man

Thomas H. Leith

In antiquity the followers of Pythagoras formed a secret cult, members of which swore not to discuss their beliefs with non-members. As much of their philosophy was concerned with abstruse numerical relationships it would have been difficult for them to be understood by outsiders anyway, and they could easily have been ridiculed by those who could not understand the basis of their thoughts. Though he lived about two thousand years after the Pythagoreans, Nicolas Copernicus would have been comfortable as a Pythagorean. Only after much pleading and cajoling by colleagues did Copernicus allow the publication of his *Revolutions of the Heavenly Spheres*, the work that launched the Copernican Revolution. Here are some of the main points of his life, as sketched by Professor Thomas H. Leith of York University:

He was born in 1473 in Torun on the Vistula River which flows into the Baltic. His father, a merchant, died when Nicolas was 10. His uncle, a Bishop, took charge of Nicolas and his sisters and brothers. At 18 Nicholas went to the University of Cracow (he was there while Columbus discovered America) but did not take the degree since his uncle called him home in order to place him as a Canon in the Frauenburg Cathedral. With the economic decline of the whole area, and especially Torun, this was to give him security. But the Canon he was to replace died a few days too soon, in an uneven month, and the custom was for the Pope to appoint on these alternate months. Copernicus missed, but 2 years later another Canon needed replacement and the uncle got him the post.

For these two years and for the next eight years Copernicus studied in Bologna and Padua at which places his interest in astronomy grew. He could stay at the University as he had no duties at the Cathedral. At 30 he took a Doctor's degree in Canon law from the University of Ferrara which was possible in those days even though he hadn't studied there. In 1506, at 33, he went to his uncle's residence castle at Heilsberg for the next 6 years, ostensibly as house physician (he had studied some medicine in Padua). While there he wrote in the *Commentary* an outline of his astronomical ideas but it was not published until the late 19th century.

In 1512 the uncle died and Copernicus took up his duties at the Frauenburg Cathedral. Here he had considerable leisure and a good income. About 1530 he finished the manuscript of the *Revolutions of the Heavenly Spheres* (early drafts began about 1515) and locked it away, making only occasional corrections later. In his 30 years there he wrote little else of import but did publish a translation of a Byzantine poet. Much of his time was involved in practicing medicine and in State affairs. The *Commentary* had become known as it was privately circulated and his views were widely known despite his failure to publish or teach. It earned him a request to help the Lateran Council reform the calendar (it occurred in 1582, but he refused, saying the motion of the Sun and Moon weren't then well enough known). We know too that the personal secretary of the pope lectured on his system and that a cardinal requested earnestly that he publish.

Copernicus hesitated to publish, however, for a variety of reasons:

1. Most of the data which he was using were known to be in error or even inconsistent. He introduced about 27 revisions of his own but even these were cruder than the observations of the Alexandrians. Consequently, the model built to fit the data was, in detail, likely to be overly complex and unsatisfactory.
2. The Reformation was bringing increasing Catholic reaction and a fear of any further departure from traditional belief made it increasingly difficult to argue that his system

was, at least in its broad principles, truer than any older system.

3. He was quite tolerant of Protestants and had friends among them. This made his own position rather difficult and his views would be, as a consequence, somewhat suspect.
4. He had an intense fear of ridicule by the uninformed and wished only the learned to debate his ideas. Indeed a play was produced near Frauenburg which did include such ridicule. There are signs that Copernicus was rather Pythagorean, that is, that he believed in keeping technical questions within the "cult" able to analyze them on their merits.
5. He almost surely believed that his views were true in principle even if dubious in detail. Consequently, while some found the views useful in improving the calendar and found the model an improved hypothesis over older ones, the difficulty with calling it true was another matter. If it were true, what then of the literal interpretation of certain Biblical texts held traditionally by most theologians? The great Lutheran scholar Melanchthon pointed this out in a book published six years after Copernicus died. Luther is also quoted as having criticized Copernicus. Catholics likewise recognized this difficulty and the *Revolutions* was placed on the Index of Forbidden Books 73 years after the book was published.

Then a young admirer, a Lutheran mathematician by the name of Rheticus, saw the manuscript of the Revolution and he and Copernicus' bishop in 1537 together tried to talk him into printing it. They failed, but Rheticus was allowed to publish the theme of the manuscript without using Copernicus' name. It repeatedly takes care to point out that Copernicus (called 'my master') goes against the ancients only when facts force him to: likely less to reassure Luther and Melanchton than to preserve Copernicus' peace of mind. The book was a masterpiece of abstracting masses of tables and complex charts and took only 10 weeks. It attracted some attention and Copernicus was urged from all sides to publish the original manuscript. Rheticus left for an important post in Leipzig in 1542, but by then he had convinced Copernicus that if he didn't publish he'd be ridiculed far worse now than if he did. Rheticus had copied the entire manuscript correcting some errors and obtained permission to publish it in Lutheran Nuremberg—so that the added weight of the Lutheran Duke of Prussia there could add his support to it in the face of Luther's and Melanchton's opposition. The book was still in press when Rheticus left for Leipzig, and Osiander, a leading Lutheran theologian, was asked to write the anonymous preface to placate the theologians and the Aristotelians whose ridicule Copernicus feared. It claimed the ideas were hypotheses (as in Ptolemy they were merely a calculating device); certainly the Copernican scheme was as good as or better than the ancient ones, but astronomy couldn't give certainty. Kepler (66 years later) reported its true author but many thought that Copernicus had written it.

Copernicus died in his tower, where he had pretty well hidden for so long, after a cerebral haemorrhage five months earlier, on May 24, 1543.

Thomas H. Leith, *The Nature and Growth of Science*, 64-66 (Toronto: Wall & Thompson, 1989).

Questions and Topics for Discussion and Writing

1. Was Copernicus justified in withholding publication for so long?
2. Comment on the breadth of Copernicus's education.

Thomas H. Leith

The Starry Messenger

Galileo Galilei

In the following excerpt from *The Starry Messenger*, Galileo announces the discoveries he made about the surface of the moon with the aid of the newly invented telescope. Elsewhere in *The Starry Messenger*, Galileo reported that the mysterious "Milky Way" was indeed a vast conglomeration of stars which could not be distinguished with the naked eye, and he reported his discovery of four satellites of Jupiter. These and other astronomical discoveries buttressed Galileo's arguments that the heavens were not unlike the Earth and that the Earth could quite logically move around the sun.

Great indeed are the things which in this brief treatise I propose for observation and consideration by all students of nature. I say great, because of the excellence of the subject itself, the entirely unexpected and novel character of these things, and finally because of the instrument by means of which they have been revealed to our senses.

Surely it is a great thing to increase the numerous host of fixed stars previously visible to the unaided vision, adding countless more which have never before been seen, exposing these plainly to the eye in numbers ten times exceeding the old and familiar stars.

It is a very beautiful thing, and most gratifying to the sight, to behold the body of the moon, distant from us almost sixty earthly radii, as if it were no farther away than two such measures—so that its diameter appears almost thirty times larger, its surface nearly nine hundred times, and its volume twenty-seven thousand times as large as when viewed with the naked eye. In this way one may learn with all the certainty of sense evidence that the moon is not robed in a smooth and polished surface but is in fact rough and uneven, covered everywhere, just like the earth's surface, with huge prominences, deep valleys, and chasms.

Again, it seems to me a matter of no small importance to have ended the dispute about the Milky Way by making its nature manifest to the very senses as well as to the intellect. Similarly it will be a pleasant and elegant thing to demonstrate that the nature of those stars which astronomers have previously called "nebulous" is far different from what has been believed hitherto. But what surpasses all wonders by far, and what particularly moves us to seek the attention of all astronomers and philosophers, is the discovery of four wandering stars not known or observed by any man before us. Like Venus and Mercury, which have their own periods about the sun, these have theirs about a certain star that is conspicuous among those already known, which they sometimes precede and sometimes follow, without ever departing from it beyond certain limits. All these facts were discovered and observed by me not many days ago with the aid of a spyglass which I devised, after first being illuminated by divine grace. Perhaps other things, still more remarkable, will in time be discovered by me or by other observers with the aid of such an instrument, the form and construction of which I shall first briefly explain, as well as the occasion of its having been devised. Afterwards I shall relate the story of the observations I have made.

About ten months ago a report reached my ears that a certain Fleming had constructed a spyglass by means of which visible objects, though very distant from the eye of the observer, were distinctly seen as if nearby. Of this truly remarkable effect several experiences were related, to which some persons gave credence while others denied them. A few days later the report was confirmed to me in a letter from a noble Frenchman at Paris, Jacques Badovere, which caused me to apply myself wholeheartedly to inquire into the means by which I might arrive at the invention of a similar instrument. This

I did shortly afterwards, my basis being the theory of refraction. First I prepared a tube of lead, at the ends of which I fitted two glass lenses, both plane on one side while on the other side one was spherically convex and the other concave. Then placing my eye near the concave lens I perceived objects satisfactorily large and near, for they appeared three times closer and nine times larger than when seen with the naked eye alone. Next I constructed another one, more accurate, which represented objects as enlarged more than sixty times. finally, sparing neither labor nor expense, I succeeded in constructing for myself so excellent an instrument that objects seen by means of it appeared nearly one thousand times larger and over thirty times closer than when regarded with our natural vision.

It would be superfluous to enumerate the number and importance of the advantages of such an instrument at sea as well as on land. But forsaking terrestrial observations, I turned to celestial ones, and first I saw the moon from as near at hand as if it were scarcely two terrestrial radii away. After that I observed often with wondering delight both the planets and the fixed stars, and since I saw these latter to be very crowded, I began to seek (and eventually found) a method by which I might measure their distances apart....

Now let us review the observations made during the past two months, once more inviting the attention of all who are eager for true philosophy to the first steps of such important contemplations. Let us speak first of that surface of the moon which faces us. For greater clarity I distinguish two parts of this surface, a lighter and a darker; the lighter part seems to surround and to pervade the whole hemisphere, while the darker part discolors the moon's surface like a kind of cloud, and makes it appear covered with spots. Now those spots which are fairly dark and rather large are plain to everyone and have been seen throughout the ages; these I shall call the "large" or "ancient" spots, distinguishing them from others that are smaller in size but so numerous as to occur all over the lunar surface, and especially the lighter part. The latter spots had never been seen by anyone before me. From observations of these spots repeated many times I have been led to the opinion and conviction that the surface of the moon is not smooth, uniform, and precisely spherical as a great number of philosophers believe it (and the other heavenly bodies) to be, but is uneven, rough, and full of cavities and prominences, being not unlike the face of the earth, relieved by chains of mountains and deep valleys. The things I have seen by which I was enabled to draw this conclusion are as follows.

On the fourth or fifth day after new moon, when the moon is seen with brilliant horns, the boundary which divides the dark part from the light does not extend uniformly in an oval line as would happen on a perfectly spherical solid, but traces out an uneven, rough, and very wavy line as shown in the figure below. Indeed, many luminous excrescences extend beyond the boundary into the darker portion, while on the other hand some dark patches invade the illuminated part. Moreover a great quantity of small blackish spots, entirely separated from the dark region, are scattered almost all over the area illuminated by the sun with the exception only of that part which is occupied by the large and ancient spots. Let us note, however, that the said small spots always agree in having their blackened parts directed toward the sun, while on the side opposite the sun they are crowned with bright contours, like shining summits. there is a similar sight on earth about sunrise, when we behold the valleys not yet flooded with light though the mountains surrounding them are already ablaze with glowing splendor on the side opposite the sun. And just as the shadows in the hollows on earth diminish in size as the sun rises higher, so these spots on the moon lose their blackness as the illuminated region grows larger and larger.

Again, not only are the boundaries of shadow and light in the moon seen to be uneven and wavy, but still more astonishingly many bright points appear within the darkened portion of the moon, completely divided and separated from the illuminated part and at a considerable distance from it. After a time these gradually increase in size and brightness, and an hour or two later they become joined with the rest of the lighted part which has now increased in size. Meanwhile more and more peaks shoot up as if sprouting now here, now there, lighting up within the shadowed portion; these become larger, and finally they too are united with that same luminous surface which extends ever further. An illustration of this is to be seen in the figure above. And on the earth, before the rising of the sun, are not the

FIGURE 1.12: The moon, on the fourth or fifth day after new moon.

highest peaks of the mountains illuminated by the sun's rays while the plains remain in shadow? Does not the light go on spreading while the larger central parts of those mountains are becoming illuminated? And when the sun has finally risen, does not the illumination of plains and hills finally become one? But on the moon the variety of elevations and depressions appears to surpass in every way the roughness of the terrestrial surface, as we shall demonstrate further on.

At present I cannot pass over in silence something worthy of consideration which I observed when the moon was approaching first quarter, as shown in the previous figure. Into the luminous part there extended a great dark gulf in the neighborhood of the lower cusp. When I had observed it for a long time and had seen it completely dark, a bright peak began to emerge, a little below its center, after about two hours. Gradually growing, this presented itself in a triangular shape, remaining completely detached and separated from the lighted surface. Around it three other small points soon began to shine, and finally, when the moon was about to set, this triangular shape (which had meanwhile become more widely extended) joined with the rest of the illuminated region and suddenly burst into the gulf of shadow like a vast promontory of light, surrounded still by the three bright peaks already mentioned. Beyond the ends of the cusps, both above and below, certain bright points emerged which were quite detached from the remaining lighted part, as may be seen depicted in the same figure. There were also a great number of dark spots in both the horns, especially in the lower one; those nearest the boundary of light and shadow appeared larger and darker, while those more distant from the boundary were not so dark and distinct. But in all cases, as we have mentioned earlier, the blackish portion of each spot is turned toward the source of the sun's radiance, while a bright rim surrounds the spot on the side away from the sun in the direction of the shadowy region of the moon. This part of the moon's surface, where it is spotted as the tail of a peacock is sprinkled with azure eyes, resembles those glass vases which have been plunged while still hot into cold water and have thus acquired a crackled and wavy surface from which they receive their common name of "ice-cups."

As to the large lunar spots, these are not seen to be broken in the above manner and full of cavities and prominences; rather, they are even and uniform, and brighter patches crop up only here and there. Hence if anyone wished to revive the old Pythagorean opinion that the moon is like another earth, its brighter part might very fitly represent the surface of the land and its darker region that of the water. I have never doubted that if our globe were seen from afar when flooded with sunlight, the land regions would appear brighter and the watery regions darker. The large spots in the moon are also seen to be less elevated than the brighter tracts, for whether the moon is waxing or waning there are always seen, here and there along its boundary of light and shadow, certain ridges of brighter hue around the large spots (and we have attended to this in preparing the diagrams); the edges of these spots are not only lower, but also more uniform, being uninterrupted by peaks or ruggedness.

Near the large spots the brighter part stands out particularly in such a way that before first quarter and toward last quarter, in the vicinity of a certain spot in the upper (or northern) region of the moon, some vast prominences arise both above and below as shown in the figures reproduced below. Before last quarter this same spot is seen to be walled about with certain blacker contours which, like the loftiest mountaintops, appear darker on the side away from the sun and brighter on that which faces the sun. (This is the opposite of what happens in the cavities, for there the part away from the sun appears brilliant, while that which is turned toward the sun is

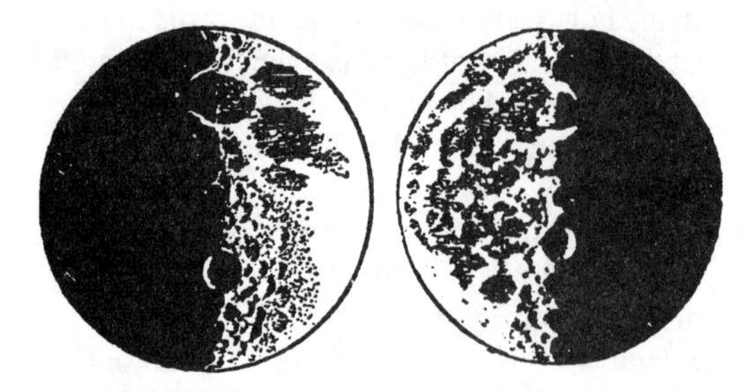

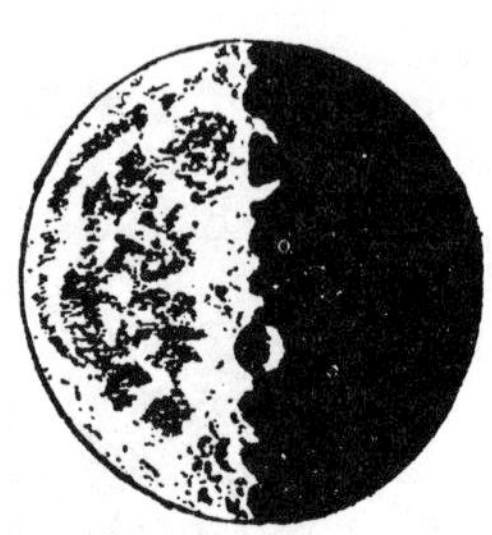

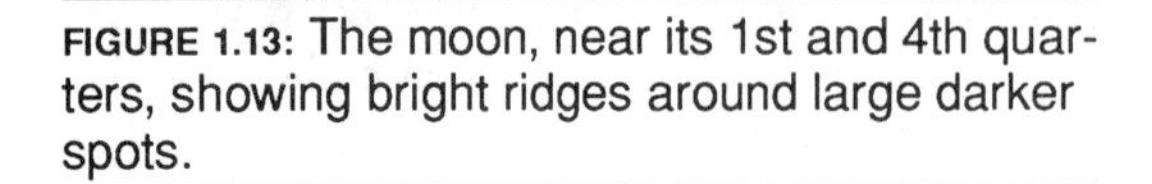

FIGURE 1.13: The moon, near its 1st and 4th quarters, showing bright ridges around large darker spots.

FIGURE 1.14: The moon, just before the last quarter, showing in the northern region a large spot "walled in" with black contours (left picture) and, after a time, the "wall" in the dark region emerges as bright ridges (right picture). Also shown is the large cavity in the southern region with the appearance of "Bohemia" viewed from the sky.

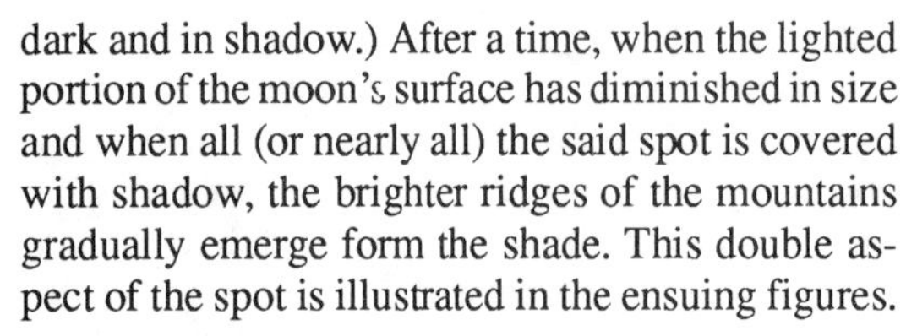

dark and in shadow.) After a time, when the lighted portion of the moon's surface has diminished in size and when all (or nearly all) the said spot is covered with shadow, the brighter ridges of the mountains gradually emerge form the shade. This double aspect of the spot is illustrated in the ensuing figures.

There is another thing which I must not omit, for I beheld it not without a certain wonder; this is that almost in the center of the moon there is a cavity larger than all the rest, and perfectly round in shape. I have observed it near both first and last quarters, and have tried to represent it as correctly as possible in the second of the above figures. As to light and shade, it offers the same appearance as would a region like Bohemia[1] if that were enclosed on all sides by very lofty mountains arranged exactly in a circle. Indeed, this area on the moon is surrounded by such enormous peaks that the bounding edge adjacent to the dark portion of the moon is seen to be bathed in sunlight before the boundary of light and shadow reaches halfway across the same space. As in other spots, its shaded portion faces the sun while its lighted part is toward the dark side of the moon; and for a third time I draw attention to this as a very cogent proof of the ruggedness and unevenness that pervades all the bright region of the moon. Of these spots, moreover, those are always darkest which touch the boundary line between light and shadow, while those farther off appear both smaller and less dark, so that when the moon ultimately becomes full (at opposition[2] to the sun), the shade of the cavities is distinguished from the light of the places in relief by a subdued and very tenuous separation.[3]

The things we have reviewed are to be seen in the brighter region of the moon. In the large spots, no such contrast of depressions and pominences is

[1] This casual comparison between a part of the moon and a specific region on earth was later the basis of much trouble for Galileo.... Even in antiquity the idea that the moon (or any other heavenly body) was of the same nature as the earth had been dangerous to hold. The Athenians banished the philosopher Anaxagoras for teaching such notions, and charged Socrates with blasphemy for repeating them.

[2] Opposition of the sun and moon occurs when they are in line with the earth between them (full moon, or lunar eclipse); conjunction, when they are in line on the same side of the earth (new moon, or eclipse of the sun).

[3] Galileo was on to more than he realized. The "cavity" he described was a giant crater formed by the impact of a huge meteorite. His comparison of its appearance to Bohemia (which he had never seen from the ground, let alone from the air) was to establish that the moon was rocky and uneven, like parts of the earth. Bohemia was ringed by mountains and would, Galileo surmised, look much the same through a telescope from a comparable distance as the "cavity" on the moon did. Indeed it does, as we now know. But more than that, some scientists are now of the opinion that the "Bohemian Plateau" in western Czechoslovakia is itself a crater formed by the impact of a huge meteorite. See Robert Kunzig, "Back to Bohemia," *Discover* 10 (June 1989): 22–23 *[Editor's note.]*

perceived as that which we are compelled to recognize in the brighter parts by the changes of aspect that occur under varying illumination by the sun's rays throughout the multiplicity of positions from which the latter reach the moon. In the large spots there exist some holes rather darker than the rest, as we have shown in the illustrations. Yet these present always the same appearance, and their darkness is neither intensified nor diminished, although with some minute difference they appear sometimes a little more shaded and sometimes a little lighter according as the rays of the sun fall on them more or less obliquely. Moreover, they join with the neighboring regions of the spots in a gentle linkage, the boundaries mixing and mingling. It is quite different with the spots which occupy the brighter surface of the moon; these, like precipitous crags having rough and jagged peaks, stand out starkly in sharp contrasts of light and shade. And inside the large spots there are observed certain other zones that are brighter, some of them very bright indeed. Still, both these and the darker parts present always the same appearance; there is no change either of shape or of light and shadow; hence one may affirm beyond any doubt that they owe their appearance to some real dissimilarity of parts. They cannot be attributed merely to irregularity of shapes, wherein shadows move in consequence of varied illuminations from the sun, as indeed is the case with the other, smaller, spots which occupy the brighter part of the moon and which change, grow, shrink, or disappear from one day to the next, as owing their origin only to shadows of prominences.

From *Discoveries and Opinions of Galileo,* trans. and ed. by Stillman Drake, 27–38 (New York: Doubleday Anchor, 1957).

Questions and Topics for Discussion and Writing

1. Why was it so important for Galileo to demonstrate that the moon was not a perfect sphere?
2. Referring to Galileo's drawings of the appearance of the moon at different times, explain how these drawings indicate aspects of the surface of the moon. What, in particular, does Galileo show by comparing views of the moon in its first and last quarters?
3. Galileo repeatedly compared his observations of shadows and reflections on the moon to similar observations on earth. Why was this dangerous and provocative for him to do?

* The Sentence of Galileo by the Inquisition, 1633

The Inquisition

In 1632, Galileo published his *Dialogue Concerning the Two Chief World Systems* in which his three interlocutors discuss the merits of the Copernican system versus the Aristotelian-Ptolemaic system and in which clearly the Aristotelian-Ptolemaic earth-centered system gets much the worst of the argument. Within five months the book was banned. Galileo, in poor health, was called before the Inquisition, tried and convicted of suspected heresy. The possible punishments were severe including torture and burning at the stake. Under the circumstances Galileo's sentence was gentle, ordering him to recant and placing him under house arrest for the rest of his life. This is the decree of the Inquisition.

Whereas you, Galileo, son of the late Vincenzo Galilei, Florentine, aged seventy years, were in the year 1615 denounced to this Holy Office for holding as true the false doctrine taught by some that the Sun is the centre of the world and immovable and that the Earth moves, and also with a diurnal motion; for having disciples to whom you taught the same doctrine; for holding correspondence with certain mathematicians of Germany concerning the same: for having printed certain letters, entitled "On the Sunspots," wherein you developed the same doctrine as true; and for replying to the objections from the Holy Scriptures, which from time to time were urged against it, by glossing the said Scriptures according to your own meaning: and whereas there was thereupon produced the copy of a document in the form of a letter, purporting to be written by you to one formerly your disciple, and in this divers propositions are set forth, following the position of Copernicus, which are contrary to the true sense and authority of Holy Scripture:

This Holy Tribunal being therefore of intention to proceed against the disorder and mischief thence resulting, which went on increasing to the prejudice of the Holy Faith, by command of His Holiness and of the Most Eminent Lords Cardinals of this supreme and universal Inquisition, the two propositions of the stability of the Sun and the motion of the Earth were by the theological Qualifiers qualified as follows:

The proposition that the Sun is the centre of the world and does not move from its place is absurd and false philosophically and formally heretical, because it is expressly contrary to the Holy Scripture.

The proposition that the Earth is not the centre of the world and immovable but that it moves, and also with a diurnal motion, is equally absurd and false philosophically and theologically considered at least erroneous in faith.

But whereas it was desired at that time to deal leniently with you, it was decreed at the Holy Congregation held before his Holiness on the twenty-fifth of February, 1616, that his Eminence the Lord Cardinal Bellarmine should order you to abandon altogether the said false doctrine and, in the event of your refusal, that an injunction should be imposed upon you by the Commissary of the Holy Office to give up the said doctrine and not to teach it to others, not to defend it, nor even discuss it; and failing acquiescence in this injunction, that you should be imprisoned. And in execution of this decree, on the following day, at the Palace, and in the presence of his Eminence, the said Lord Cardinal Bellarmine, after being gently admonished by the said Lord Cardinal, the command was enjoined upon you by the Father Commissary of the Holy Office of that time, before a notary and witnesses, that you were altogether to abandon the said false opinion and not in future to hold or defend or teach it in any way whatsoever, neither verbally nor in writing; and upon your promising to obey, you were dismissed.

And, in order that a doctrine so pernicious might be wholly rooted out and not insinuate itself further to the grave prejudice of Catholic truth, a decree was issued by the Holy Congregation of the Index prohibiting the books which treat of this doctrine and declaring the doctrine itself to be false and wholly contrary to the sacred and divine Scripture.

And whereas a book appeared here recently, printed last year at Florence, the title of which shows that you were the author, this title being: "Dialogue of Galileo Galilei on the Great World Systems"; and whereas the Holy Congregation was afterward informed that through the publication of the said book the false opinion of the motion of the Earth and the stability of the Sun was daily gaining ground, the said book was taken into careful consideration, and in it there was discovered a patent violation of the aforesaid injunction that had been imposed upon you, for in this book you have defended the said opinion previously condemned and to your face declared to be so, although in the said book you strive by various devices to produce the impression that you leave it undecided, and in express terms as probable: which, however, is a most grievous error, as an opinion can in no wise be probable which has been declared and defined to be contrary to divine Scripture.

Therefore by our order you were cited before this Holy Office, where, being examined upon your oath, you acknowledged the book to be written and published by you. You confessed that you began to write the said book about ten or twelve years ago, after the command had been imposed upon you as above; that you requested license to print it without, however, intimating to those who granted you this license that you had been commanded not to hold, defend, or teach the doctrine in question in any way whatever.

You likewise confessed that the writing of the said book is in many places drawn up in such a form that the reader might fancy that the arguments brought forward on the false side are calculated by their cogency to compel conviction rather than to be easy of refutation, excusing yourself for having fallen into an error, as you alleged, so foreign to your intention, by the fact that you had written in dialogue and by the natural complacency that every man feels in regard to his own subtleties and in showing himself more clever than the generality of men in devising, even on behalf of false propositions, ingenious and plausible arguments.

And, a suitable term having been assigned to you to prepare your defense, you produced a certificate in the handwriting of his Eminence the Lord Cardinal Bellarmine, procured by you, as you asserted, in order to defend yourself against the calumnies of your enemies, who charged that you had abjured and had been punished by the Holy Office, in which certificate it is declared that you had not abjured and had not been punished but only that the declaration made by His Holiness and published by the Holy Congregation of the Index had been announced to you, wherein it is declared that the doctrine of the motion of the Earth and the stability of the Sun is contrary to the Holy Scriptures and therefore cannot be defended or held. And, as in this certificate there is no mention of the two articles of the injunction, namely, the order not "to teach" and "in any way," you represented that we ought to believe that in the course of fourteen or sixteen years you had lost all memory of them and that this was why you said nothing of the injunction when you requested permission to print your book. And all this you urged not by way of excuse for your error but that it might be set down to a vainglorious ambition rather than to malice. But this certificate produced by you in your defense has only aggravated your delinquency, since, although it is there stated that said opinion is contrary to Holy Scripture, you have nevertheless dared to discuss and defend it and to argue its probability; nor does the license artfully and cunningly extorted by you avail you anything, since you did not notify the command imposed upon you.

And whereas it appeared to us that you had not stated the full truth with regard to your intention, we thought it necessary to subject you to a rigorous examination at which (without prejudice, however, to the matters confessed by you and set forth as above with regard to your said intention) you answered like a good Catholic. Therefore, having seen and maturely considered the merits of this your cause, together with your confessions and excuses above-mentioned, and all that ought justly to be seen and considered, we have arrived at the underwritten final sentence against you:

Invoking, therefore, the most holy name of our Lord Jesus Christ and of His most glorious Mother, ever Virgin Mary, by this our final sentence, which sitting in judgment, with the counsel and advice of the Reverend Masters of sacred theology and Doctors of both Laws, our assessors, we deliver in these writings, in the cause and causes at present before us between the Magnificent Carlo Sinceri, Doctor of both Laws, Proctor Fiscal of this Holy Office, of the one part, and you Galileo Galilei, the defendant, here present, examined, tried, and confessed as shown above, of the other part—

We say, pronounce, sentence, and declare that you, the said Galileo, by reason of the matters adduced in trial, and by you confessed as above, have rendered yourself in the judgment of this Holy Office vehemently suspected of heresy, namely, of having believed and held the doctrine — which is false and contrary to the sacred and divine Scriptures — that the Sun is the centre of the world and does not move from east to west and that the Earth moves and is not the centre of the world; and that an opinion may be held and defended as probable after it has been declared and defined to be contrary to the Holy Scripture; and that consequently you have incurred all the censures and penalties imposed and promulgated in the sacred canons and other constitutions, general and particular, against such delinquents. From which we are content that you be absolved, provided that, first, with a sincere heart and unfeigned faith, you abjure, curse, and detest before us the aforesaid errors and heresies and every other error and heresy contrary to the Catholic and Apostolic Roman Church in the form to be prescribed by us for you.

And, in order that this your grave and pernicious error and transgression may not remain altogether unpunished and that you may be more cautious in the future and an example to others that they may abstain from similar delinquencies, we ordain that the book of the "Dialogue of Galileo Galilei" be prohibited by public edict.

We condemn you to the formal prison of this Holy Office during our pleasure, and by way of salutary penance we enjoin that for three years to come you repeat once a week the seven penitential Psalms. Reserving to ourselves liberty to moderate, commute, or take off, in whole or in part, the aforesaid penalties and penance.

And so we say, pronounce, sentence, declare, ordain, and reserve in this and in any other better way and form which we can and may rightfully employ.

Quoted in Giorgio de Santillana, *The Crime of Galileo, 306-310 (Chicago: The University of Chicago Press, 1955).*

Questions and Topics for Discussion and Writing

1. Why was the Inquisition so upset about Galileo's views? What difference did it make to them what people thought about the motions of the planets and the earth?
2. Was this sentence the inevitable result of Galileo's activities? Could there have been a different resolution of Galileo's disagreement with the Church? If so, what could that have been and why? If not, why not?

Newton: The Man

Thomas H. Leith

It was Isaac Newton who brought together the pieces of the new scientific theories and discoveries about the physical world made by Copernicus, Kepler, Galileo, Descartes, and others. His great synthesis, which we still call Newtonian physics, completely replaced Aristotle's physics as fundamental scientific laws. The key points: the laws of inertia, acceleration, action and reaction, and the principle of universal gravitation are the foundation of physical science. Also Newton invented the calculus and was the first to analyze the composition of white light. These are the main achievements for which he is remembered today. But this is but a tiny fraction of Newton's work, the vast bulk of which was not about science at all and most of which was never published.

Amazing as Newton's scientific achievements were, his personal life is equally amazing. He certainly ranks as one of the great eccentrics in history. Here, Professor Leith sketches some of the main events in Newton's life.

Nature, and Nature's laws lay hid in night:
God said, *Let Newton be!* and all was light.
Alexander Pope,

Newton was born at Woolsthorpe within a year of Galileo's death—1642—in Lincolnshire, England. His father was a farmer, who died before Newton's birth. At 12 he moved away to school in Grantham where he lived with a druggist who taught him skills in the manual arts. Thus he built windmills to scale, water clocks, sundials, kites, and lanterns. These were largely copied from books which we still have, just as we have his notes from them. He did not show his genius early (nothing like Pascal who at 16 had made major discoveries in geometry; or Hamilton who at 14 spoke 8 languages, at 16 wrote a famous paper in mathematics, around 20 was Royal Astronomer of Ireland and was the leading mathematician of his age; or Maxwell whose first paper in mathematics was published at 15), though we know he stood first in his tiny school. At 18 he went to Trinity College at Cambridge where he worked his way doing odd jobs. We know little of him until his third year when he met the holder of a mathematics chair at the University and who turned his interests further to mathematics and to optics.

In 1665 he took his B.A. without any great distinction. The University closed because of the plague and he returned home and spent much of his 23rd and 24th years there. He tells us little of this period but indications are that the period was of great import in his life. At home he seems to have had much leisure time and spent it in intense concentrated thought for hours on end. Here he developed his law of gravitational attraction and likely his famous 3 laws of motion, not published until over 20 years later. Here too he made major discoveries in the differential and integral calculus and he reports that he solved areas between curves and lines to many decimals *in his head*. (It was 4 years later in 1669 at 26 years of age that he sent his first result to his mathematics professor friend, but they weren't published for 30 more years. This reticence to publish led to disputes whether he or Leibniz had discovered the calculus first and as to possible plagiarism). Here, too, he bought his prism to study light and colours.

In 1667 he returned to Cambridge as a minor fellow, becoming a major fellow the next year. Two other major fellows were injured falling drunk downstairs and another went insane. When his mathematics friend resigned the next year, Newton became the Lucasian Professor of Mathematics (the position held recently by the great Dirac and now held by Stephen W. Hawking). One of his friends at this time said of him that he was "a man of quite exceptional ability and singular skill." At this time he built a number of lenses and discovered chromatic aberration. To get around it he built a reflecting telescope (an early form of those in many observatories today) which aroused great interest in Cambridge and at the Royal Society in London. Indeed, he gave it to the latter and they still have it. Feeling that was too easy a task to fit the honour, he wrote his first published paper in 1672 (at 30) on light and prisms. Robert Hooke reviewed it and suggested some criticisms (it opposed certain ingenious ideas of his own) which Newton listened to politely. Later these two great experimenters and theoreticians were prodded into a nasty dispute by an enemy of Hooke's. Huygens criticized Newton also, partly from lack of understanding of Newton's point that he wanted to measure the behaviour of colors and not argue about their *nature*. Several others were so obnoxious that Newton swore in 1676 to thenceforth leave science only for his private satisfaction. Fortunately he relented.

Before this, in 1674, Hooke wrote him about planetary motion and renewed his interest in it. In 1684 Christopher Wren asked Hooke to prove a certain theory Hooke had proposed that the forces pulling the planets out of straight paths into curves around the sun obeyed an inverse square law. While Hooke was correct in his guess he couldn't prove it. Later Wren and Halley (of comet fame) reversed the question and asked Newton what kind of curve such a law would give. He replied at once—an ellipse. They asked in amazement how he knew. He said he had calculated it but lost his papers. He did the work again, however and in two different ways! Halley saw a major discovery in it and asked Newton for a thorough treatise on his work.

As a result, in 1686 Newton sent the first book, of what was to be three to the Royal Society. The minutes of the meeting record its presentation as *Philosophiae Naturales Principia Mathematica*—among other things a demonstration of the Copernican hypothesis and of Kepler's laws for elliptical orbits, using the inverse square law of attraction toward the sun. Within 18 months all three books were finished and publishing was begun. Just earlier, Hooke had noticed that his suggestions of 6 years previous was used by Newton without acknowledgement. Of course he didn't know Newton had had the ideas as long as 15–20 years before, though it is true that his letter in 1674 had renewed Newton's interest. Newton took offense when he was told of Hooke's chagrin, painted worse than it was, likely because of his prior experience with Hooke and threatened not to finish the third book just as he had threatened to leave science earlier. Eventually Newton saw Hooke had been misrepresented and was asking only for justice from the Society, and Newton mentioned Hooke in the *Principia* when it was published.

We are told that at this time Newton was very absent minded—indeed accounts suggest he always was. He'd forget to eat or he'd go out for wine for a party and forget to return. He'd dash up to his rooms to write an idea while fresh in his mind—standing beside a chair without noticing it. The *Principia* came out in 1687, financed by Halley (the Royal Society was too poor), and it sold for six shillings. It is much like Euclid's "Elements"—the laws are derived from geometry. He could have done it far more easily using the calculus but he didn't. Just why isn't clear. He does tell us that it was made abstruse purposely so that only mathematicians could read it and he wouldn't have to deal with questions from lesser mortals. He suggests these lesser mortals read Book Three which is a more popular presentation.

Scientists immediately saw the great value of the work and we have extant glowing reviews of the work calling Newton the greatest living mathematician. However, it wasn't used at once. Six years later Cambridge was still using the old Cartesian arguments. Part of the reason lay in its difficulty and one famous mathematician reported that he had to tear it up and carry a few pages at a time so he could digest it.

In 1689 he went to Parliament as the representative of Cambridge University. In London he met Pepys, John Locke, and many other great figures of the time. Later, Newton sought one administrative post after another. Failing, he neared a nervous breakdown and accused his friends of various evils toward him. His interest in science at this time also lagged or resulted in quarrels. However, in 1696 he was made Warden of the Mint and 3 years later the Master (or Head) with a good salary and considerable leisure time. (There are stories of a love affair but all evidence points to it being untrue. Indeed, we have no record of Newton's ever having been in love.) He turned out to be a good administrator but did little science, though we are told that Jean Bernoulli (the great Swiss mathematician) set a problem about this time as a challenge to all European mathematicians and Newton did it in one evening.

In 1703 he became president of the Royal Society and so remained till his death in 1727. Usually the president was not a scientist and this was a great honour. In 1704 he published his *Opticks*—his great work on light—based largely upon studies long before but delayed to avoid controversy. It is coincidence that Hooke died just before it appeared. The second edition, however, has various new sections of great thought and worth—evidence he could still contribute to science when he desired. One is on polarized light and another on electrical attraction—both major ideas.

In 1705 Queen Anne knighted him *and went to Cambridge to do it*. He was now 62. Most of his time was spent in the Mint or dealing with the problems of the Royal Society (mostly quarrels). The weekly meetings today are on Thursday—because they were changed from Wednesday for Newton's convenience at the Mint. Part of his time he spent fighting on his own account, especially with an astronomer called Flamsteed. He treated him rather highhandedly to say the least and in later editions left out portions of the first edition of the *Principia* where he was mentioned. Newton had a proud side! Leibniz too had caused trouble before Newton left for London and after the publication of the *Opticks*. Most of the inciting was on Newton's part however: he accused Leibniz of plagiarism and sponsored a book against him at the Royal Society. The trouble died only with Leibniz's death in 1716.

The second edition of the *Principia* appeared in 1713 with revisions by a brilliant young friend of Newton's. The first edition was already scarce. The new preface points out that many still didn't believe it, being unconvinced by experimental and observational data. A third edition appeared in 1726 and now it was written in other than Latin for the first time. In 1711 Newton published his calculus treatise written many years before and just before this a work on algebra. Another book on his optical lectures came out after his death (he never actually did the work of preparing any of his books for publication anyway). Most of his interest at this time lay in theology and history, though he had corresponded with Locke as early as 1691 on the prophecies in Daniel. His notes contain a million and a half words on religion and theology (some of it has been only recently published) and another half million words on Alchemy. Two books on this part of his work were published, one before his death. They deal with Biblical, Greek, Egyptian, and Assyrian chronologies and with the Biblical Books of Daniel and Revelation. The latter was a concerted attempt at a correlation of prophecy with history, evidencing deep learning in church history. He said of it "it is the noblest use to which the human intellect can be applied."

In 1727 he went to London to preside at the Royal Society. The trip was too much for him and he died a month later in March—aged 85. His body lay in state and was buried in Westminster Abbey. The casket was carried by 2 dukes and 3 earls, and the Lord High Chancellor. Present at his funeral also was Voltaire, who did much to popularize his views on the Continent.

In closing let us briefly review his character.

1. He could almost intuit the solution to a mathematical problem, often unproven for many years.
2. He anticipated much later theory in varied areas.
3. His scientific contributions came in rather brief periods in a long life. One wonders what might have resulted had he sustained his attack. Indeed 75% of his time he spent with the Mint, religious matters, and alchemy.
4. He was deeply religious but was not capable of full agreement with the 39 articles of the Church of England. His works on prophecy were meant to show the stupidity of trying to foreknow but the great use of historical data in interpreting prophecy *post facto*. Allied with this were his mystical and alchemical interests. His library had many works by the chemist Boyle and by the Rosicrucians among others. Even while writing his scientific works he spent much time in the alchemical laboratory. "What his aim might be I was not able to penetrate into," a friend tells us, and his writings never make it clear. Certainly it was a great driving passion in his life but, typically, he kept the public and the printer

out of his interests. Remember, others always had to drag him into publications and apparently had not sufficient interest or understanding to lead him to publish his chemical work.

5. He could be very spiteful and ungenerous when his authority was questioned but, to his students of capability, he was most helpful. Toward truth itself, rather than toward the portion he felt he had, he was most humble. Before he died he made his famed remark "I do not know what I may appear to the world but to myself I seem to have been only like a boy, playing on the seashore, and diverting myself in now and then finding a smoother pebble or a prettier shell than ordinary, while the great ocean of truth lay all undiscovered before me." Likely his religious and alchemical interests are part of this broader view.

Thomas H. Leith, *The Nature and Growth of Science,* 113-117 (Toronto: Wall & Thompson, 1989).

Questions and Topics for Discussion and Writing

1. Comment on the different personalities of Copernicus, Galileo, and Newton, based on what you have read about each of them. Were their personalities suited to the scientific work they did?
2. Newton spent much more of his time on theology and alchemy than on the mathematics and physics for which we remember him. Do you think he would have accomplished more if he had spent more of his time on mathematics and physics? Why or why not?

* Einstein

James R. Newman

Newton's physics provided a secure foundation for the development of modern physical science. The Newtonian model of the universe is a good enough description of the world that an enormous amount of physical theory and application has been built upon it, including most of the practical applications known to us today. But the Newtonian world view is not quite satisfactory at the extremes. Newton's laws do not describe physical phenomena well at a very small scale, such as at a subatomic level, nor at very large scales, such as at astronomical distances or at speeds close to the speed of light.

Near the end of the nineteenth century, Newtonian physics had been pursued so far that some of the anomalies of large and small scale phenomena began to be noticed. Leading scientists turned their attention to these problems and began to record and analyze the unusual observations. In the first half of the twentieth century a whole new physics arose, not so much replacing Newtonian physics, but extending and enlarging it. At the subatomic level, quantum mechanics took over where Newton's laws no longer were adequate. At the level of the very large scale, relativity theory explained events that Newtonian physics could not. Both quantum mechanics and relativity theory posed new questions, not just about physics, but about the nature of knowledge and of existence itself. Thus the new physics of this century has opened a vast set of problems and issues for scientists, philosophers, and laymen to contemplate.

Though quantum mechanics is associated with several different important scientists, relativity theory is primarily the creation of one man, Albert Einstein. (Quantum mechanics is based upon a number of fundamental implications of atomic physics, many of which were inferred by Einstein. But Einstein himself could not accept some of the implications of quantum mechanics.) The following review of Einstein's work by James R. Newman gives an excellent overview of the scope, development, and importance of relativity, and some insight into the mind and person of this century's most renowned scientific genius.

To the eyes of the Man of Imagination, Nature is Imagination Itself.

William Blake

Einstein died in 1955. Some fifty years earlier, when he was twenty-six, he put forward an idea that changed the world. His idea revolutionized our conception of the physical universe; its consequences have shaken human society. Since the rise of science in the seventeenth century, only two other men, Newton and Darwin, have produced a comparable upheaval in thought.

Everybody knows that Einstein did something remarkable, but what exactly did he do? Even among educated men and women, few can answer. We are resigned to the importance of his theory, but we do not comprehend it. It is this circumstance that is largely responsible for the isolation of modern science. This is bad for us and bad for science; therefore more than curiosity is at stake in the desire to understand Einstein.

Relativity is a hard concept, prickly with mathematics. there are many popular accounts of it, a small number of which are good, but it is a mistake to expect they will carry the reader along—like a prince stretched on his palanquin. One must tramp one's own road. Nevertheless, relativity is in some respects simpler than the theory it supplanted. It makes the model of the physical world more susceptible to proof by experiment; it replaces a grandiose scheme of space and time with a more practical scheme. Newton's majestic system was worthy of the gods; Einstein's system is better suited to creatures like ourselves, with limited intelligence and weak eyes.

But relativity is radically new. It forces us to

change deeply rooted habits of thought. It requires that we free ourselves from a provincial perspective. It demands that we relinquish convictions so long held that they are synonymous with common sense, that we abandon a picture of the world that seems as natural and as obvious as that the stars are overhead. it may be that Einstein's ideas will seem easy to a generation that has grown up with them; but for our generation, as Bertrand Russell has said, a complete "imaginative reconstruction is unavoidable." Anyone who seeks to understand the world of the twentieth century must make this effort.

* * *

In 1905, while working as an examiner in the Swiss Patent Office, Einstein published in the *Annalen der Physik* a thirty-page paper with the title "On the Electrodynamics of Moving Bodies." The paper embodied a vision. Poets and prophets are not alone in their visions; a young scientist—it happens mostly to the young—may in a flash glimpse a distant peak that no one else has seen. He may never see it again, but the landscape is forever changed. The single flash suffices; he will spend his life describing what he saw, interpreting and elaborating his vision, giving new directions to other explorers. Einstein had not been a wonder child. He had not astonished his teachers or his parents. His father showed him a compass at the age of five; the event, he tells us, made a profound and lasting impression upon him. For the first time he realized that "something deeply hidden had to be behind things." When he was twelve, a book on Euclidean geometry came into his hands; this too was an unforgettable event. His "holy geometry booklet," as he called it, opened a new delight: it was possible to prove things, to attain certainty by pure thinking. He continued his studies in mathematics and he also read with "breathless attention" a six-volume popular work that comprised the entire field of the natural sciences. At the age of seventeen, having already dipped into theoretical physics, he entered the Polytechnic Institute of Zurich. By then the problem that was to culminate in his vision had begun to stir in him.

At the heart of the theory of relativity are questions connected with the velocity of light. The young Einstein began to brood about these while still a high-school student. Suppose, he asked himself, a person could run as fast as a beam of light, how would things look to him? Imagine that he could ride astride the beam, holding a mirror just in front of him. Then, like a fictional vampire, he would cast no image; for since the light and the mirror are traveling in the same direction at the same velocity, and the mirror is a little ahead, the light can never catch up to the mirror and there can be no reflection.

But this applies only to *his* mirror. Imagine a stationary observer, also equipped with a mirror, who watches the rider flashing by. Obviously the observer's mirror will catch the rider's image. In other words the optical phenomena surrounding this event are purely relative. They exist for the observer; they do not exist for the rider. This was a troublesome paradox, which flatly contradicted the accepted views of optical phenomena. We shall have to see why.

The velocity of light had long engaged the attention of physicists and astronomers. In the seventeenth century the Danish astronomer Römer discovered that light needed time for its propagation. Thereafter, increasingly accurate measurements of its velocity were made, and by the end of the nineteenth century the established opinion was that light in a vacuum always travels at a certain constant rate, about 186,000 miles a second.

But now a new problem arose. In the mechanics of Galileo and Newton, rest and uniform motion (i.e., constant velocity) are regarded as indistinguishable. Of two bodies, A and B, it can only be said that one is in motion *relative* to the other. The train glides by the platform; or the platform glides by the train. the earth approaches the fixed stars; or the stars approach the earth. There is no way of deciding which of these alternatives is true. And in the science of mechanics it makes no difference.

One of the questions, therefore, was whether, in respect of motion, light itself was like a physical body; that is, whether its motion was relativistic in the Newtonian sense, or absolute.

The wave theory of light appeared to answer this question. A wave is a progressive motion in some kind of medium; a sound wave, for example, is a movement of air particles. Light waves, it was supposed, move in an all-pervasive medium called the ether. The ether was assumed to be a subtle jelly with marvellous properties. It was colorless, odorless, without detectable features of any kind. It could penetrate all matter. It quivered in transmitting light. Also, the body of the ether as a whole was held to be stationary. To the physicist this was its most important property, for, being absolutely at rest, the ether offered a unique frame of reference for determining the velocity of light. Thus while it was hopeless to attempt to determine the absolute motion of a physical body, because one could find no absolutely stationary frame of reference against which to measure it, the attempt was not hopeless for light: the ether, it was thought, met the need.

The ether, however, did not meet the need. Its marvellous properties made it a terror for experimentalists. How could motion be measured against an ectoplasm, a substance with no more substantiality than an idea? Finally, in 1887, two American physicists, A.A. Michelson (the first American, by the way, to win the Nobel prize in physics) and E.W. Morley, rigged up a beautifully precise instrument, called an interferometer, with which they hoped to discover some evidence of the relationship between light and the hypothetical ether. If the earth moves through the ether, a beam of light traveling in the direction of the earth's motion should move faster through the ether than a beam traveling in the opposite direction. Moreover, just as one can swim across a river and back more quickly than one can swim the same distance up and down stream, it might be expected that a beam of light taking analogous paths through the ether would complete the to-and-fro leg of the journey more quickly than the up-and-down leg.

This reasoning was the basis of the Michelson-Morley experiment. With great care and ingenuity they carried out a number of trials in which they compared the velocity of a beam of light moving through the ether in the direction of the earth's motion, and another beam traveling at right angles to this motion. There was, as I have indicated, every reason to believe that these velocities would be different. Yet no difference was observed. The results were entirely negative. The light beam seemed to move at the same velocity in either direction. The possibility that the earth dragged the ether with it having been ruled out, the inquiry had come to a dead end. Perhaps there was no difference; perhaps there was no ether. The Michelson-Morley findings were a major paradox.

Various ideas were advanced to resolve it. The most imaginative of these, and also the most fantastic, was put forward by the Irish physicist G.F. FitzGerald. He suggested that since matter is electrical in essence and held together by electrical forces, it may contract in the direction of its motion as it moves through the ether. the contraction would be very small; nevertheless in the direction of motion the unit of length would be shorter. This hypothesis would explain the Michelson-Morley result. The arms of their interferometer might contract as the earth rotated; this would shorten the unit of length and cancel out the added velocity imparted to the light by the rotation of the earth. The velocities of the two beams—in the direction of the earth's motion and at right angles to it—would appear equal. FitzGerald's idea was elaborated by the famous Dutch physicist H. A. Lorentz. He put it in mathematical form and connected the contraction caused by motion with the velocity of light. According to his arithmetic the contraction was just enough to account for the negative results of the Michelson-Morley experiment. There the subject rested until Einstein took it up anew.

He knew of the Michelson-Morley findings (though there is evidence to show that they played a surprisingly small part in leading him to his own theory). He knew also of other inconsistencies in the contemporary model of the physical world. One was the slight but persistent misbehavior (by classical standards) of the planet Mercury as it moved in its orbit; it was losing time (at a trifling rate, to be sure—43 seconds of arc per century), but Newton's theory of its motion was exact and there was no way of accounting for the discrepancy. Another was the bizarre antics of electrons, which, as W. Kaufmann and J.J. Thomson discovered, increased in mass as they went faster. The question was, could these inconsistencies be overcome by patching and mending classical theories, by improvising in the vein of the astronomers who for centuries tacked on epicycles to Ptolemy's system to keep it alive? Or had the time come for a Copernican renovation?

Making his own way, Einstein turned to another aspect of the velocity problem. Velocity measurements involve time measurements, and time measurements, as he perceived, involve the concept of simultaneity. Is this concept simple and intuitively clear? No one doubted that it was, but Einstein demanded proof.

I enter my study in the morning as the clock on the wall begins to strike. Obviously these events are simultaneous. Assume, however, that on entering the study I hear the first stroke of the town-hall clock, half a mile away. It took time for the sound to reach me; therefore while the sound wave fell on my ears at the moment I entered the study, the event that produced the wave was not simultaneous with my entry.

Consider another kind of signal. I see the light from a distant star. An astronomer tells me that the image I see is not of the star as it is today, but of the star as it was the year Brutus killed Caesar. What does simultaneity mean in this case? Is my *here-now* simultaneous with the star's *there-then*? Can I speak meaningfully of the star as it was the day Joan of Arc was burned, even though ten generations will have to pass before the light emitted by the star on that day reaches the earth? How can I be sure it will ever get here? In short, is the concept of simultaneity for different places exactly equivalent to the concept for one and the same place?

Einstein soon convinced himself that the answer is no. Simultaneity, as he realized, depends on signals; the speed of light (or other signal) must therefore enter into the meaning of the concept. Not only does the separation of events in *space* becloud the issue of simultaneity in *time*, but relative motion may do so. A pair of events, which one observer pronounces simultaneous, may appear to another observer, in motion with respect to the first, to have happened at different times. In his own popular account of relativity (see Figure 1.15), Einstein gave a convincing and easy example, which showed that *any* measurement of time is a measurement with respect to a given observer. A measurement valid for one observer may not be valid for another. Indeed, the measurement is certain not to be valid if one attempts to extend it from the system where the measurement was made to a system in motion relative to the first.

We approach the consummation of Einstein's revolutionary theory. He was now aware of these facts. Measuring the velocity of light requires a time measurement. This involves a judgment of simultaneity. Simultaneity is not an absolute fact, the same for all observers. The individual observer's judgement depends on relative motion.

Einstein went further. To measure a time interval of any kind requires the use of a clock (or its equivalent). The method consists of matching a signal, marking the beginning of the interval, against a certain position of the hands of the clock, and matching another signal, marking the end of the interval, against another position of the hands. The distance between the two points of coincidence is a measure of the duration of the interval. Simultaneity is clearly involved.

The sequence does not end there. Simultaneity may also be involved in measuring distances. A passenger on a moving train who wants to measure the length of his car has no difficulty. With a yardstick he can do the job as easily as if he were measuring his room at home. Not so for a stationary observer watching the train go by. The car is moving and he cannot measure it simply by laying yardstick end on end. He must use light signals, which will tell him when the ends of the car coincide with certain arbitrary points. Therefore, problems of time arise. Suppose the thing to be measured is an electron, which is in continual motion at high speed. Light signals will enter the experiment, judgements of simultaneity will have to be made, and once again it is obvious that observers of the electron, who are in motion relative to each other, will get different results. The whole comfortable picture of reality begins to disintegrate: neither space nor time is what it seems.

The clarification of the concept of simultaneity thrust upon Einstein the task of challenging two assumptions, assumptions hedged with the divinity of Isaac Newton. "Absolute, true, and mathematical time, of itself and from is own nature, flows equably without relation to anything external...." This was Newton's sonorous definition in his great book, *Principia Mathematica*. To this definition he added the equally majestic "Absolute space, in its own nature, without relation to anything external, remains always similar and immovable." These assumptions, as Einstein saw, were magnificent but untenable. they were at the bottom of the paradoxes of contemporary physics. They had to be discarded. Absolute time and absolute space were concepts that belonged to an outworn metaphysic. They went beyond observation and experiment; indeed, they were refuted by the nasty facts. Physicists had to

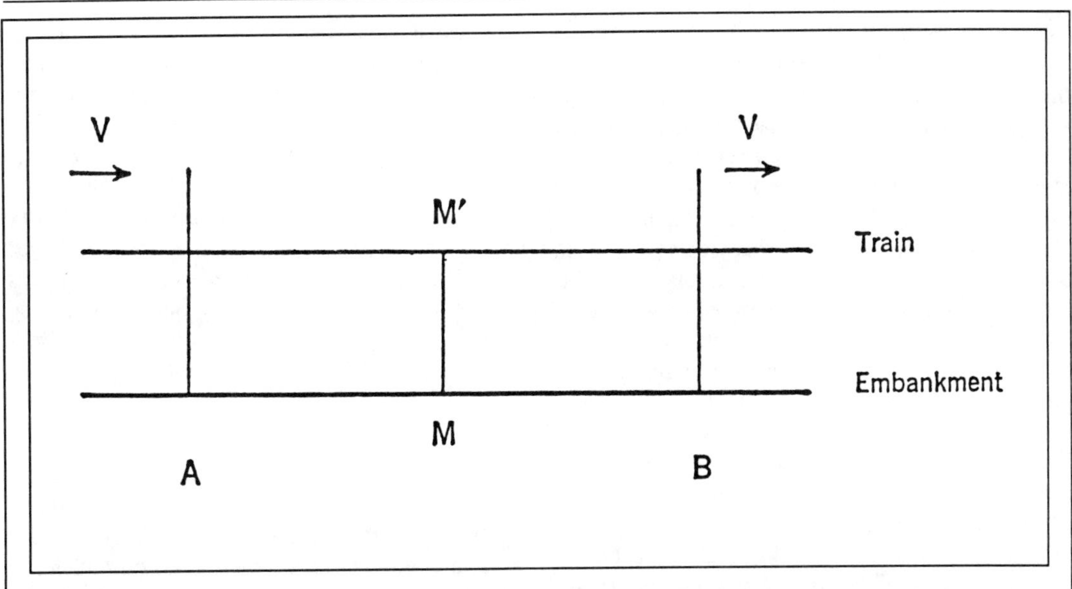

The diagram shows a long railroad train traveling along the rails with velocity V, in the direction toward the right of the page. The bottom line denotes the embankment running parallel to the rails. The letters A and B mark two places on the rails, and the letter M marks a point on the embankment directly midway between A and B. At M stands an observer, equipped with a pair of mirrors that are joined in a V and inclined at 90°. By means of this device he can observe both places A and B at the same time. We imagine two events at A and B, say two flashes of lightning, which the observer perceives in his mirror device at the same time. These he pronounces to be simultaneous, by which he means that the rays of light emitted at A and B by the lightning bolts meet at the midpoint M of the length A → B along the embankment. Now consider the moving train, and imagine a passenger seated in it. As the train proceeds along the rails, the passenger will arrive at a point *M′*, which is directly opposite M, and therefore exactly midway between the length A → B along the rails. Assume further that the passenger arrives at *M′* just when the flashes of lightning occur. We have seen that the observer at M correctly pronounces the lightning bolts to be simultaneous; the question is, will the train passenger at *M′* make the same pronouncement? It is easily shown that he will not. Obviously if the point *M′* were stationary with respect to M, the passenger would have the same impression of simultaneity of the lightning flashes as the observer on the embankment. But *M′* is not stationary; it is moving toward the right with the velocity V of the train. Therefore (considered with reference to the embankment) the passenger is moving toward the beam of light coming from B, and away from the beam coming from A. It seems clear then that he will see the beam emitted by the flash at B sooner than the beam emitted by the flash at A. Accordingly he will pronounce the flash at B as earlier in time than the flash at A.

Which of the two pronouncements is correct, the observer's or the passenger's? The answer is that each is right in his own system. The observer is right with respect to the embankment, the passenger with respect to the train. The observer may say that he alone is right because he is at rest, while the passenger is moving and his impressions are therefore distorted. To this the passenger can reply that motion does not distort the signals, and that, in any case, there is no more reason to believe that he was moving and the observer at rest than that the passenger was at rest and the observer moving.

There is nothing to choose between these views, and they can be logically reconciled only by accepting the principle that simultaneity of events is meaningful only with respect to a particular reference system; moreover, that every such system has its own particular time, and unless, as Einstein says, we are told the reference system to which the statement of time refers, a bare statement of the time of an event is meaningless.

James R. Newman

live with these facts.

In taking this bold step Einstein was animated by the spirit of certain philosophers of science. He acknowledged his debt to the Austrian positivist Ernst Mach, to the brilliant French mathematician Henri Poincaré, and of course to the great Scottish skeptic of the eighteenth century, David Hume. Hume took the cheerful view that we know nothing and that there is no such thing as a rational belief. But, he said, in our pathetic groping—and to ease our discomfort—we ought at least to pretend that experience and observation are useful guides. Einstein was not primarily a philosopher, but he saw that there were problems of physics that could not be solved without taking seriously the basic criticisms of scientific procedure that these philosophers had uttered. The central point of his approach was that *nothing* had meaning that could not be verified. A new philosophy of physics had to precede a reform of physics. Without ever explicitly formulating such a philosophy, Einstein put it into his work.

* * *

The clear way out of the Michelson-Morley paradox was to accept it. From the point of view of common sense the results were extraordinary, yet they had been verified. It was not the first time that science had had to overrule common sense. The evidence showed that the velocity of light measured by *any* observer, whether at rest or in motion relative to the light source, is the same. Einstein embodied this fact in a principle from which a satisfactory theory of the interaction between the motion of bodies and the propagation of light could be derived. This principle, or first postulate, of his Special Theory of Relativity states that *the velocity of light in space is a constant of nature, unaffected by the motion of the observer or of the source of the light.*

The hypothesis of the ether thus became unnecessary. One did not have to try to measure the velocity of light against an imaginary frame of reference, for the plain reason that whenever light is measured against *any* frame of reference its velocity is the same. Why then conjure up ethereal jellies? The ether simply lost its reason for being.

A second postulate was needed. Newtonian relativity applied to the motion of material bodies; but light waves, as I mentioned earlier, were thought not to be governed by this principle. Einstein pierced the dilemma in a stroke. He simply extended Newtonian relativity to include optical phenomena. The second postulate says: *In any experiment involving mechanical or optical phenomena it makes no difference whether the laboratory where the experiment is being performed is at rest or in uniform motion; the results of the experiments will be the same.* More generally, one cannot by any method distinguish between rest and uniform motion, except in relation to each other.

* * *

Is that all there is to the Special Theory of Relativity? The postulates are deceptively simple. Moreover, to the sharp-eyed reader they may appear to contradict each other. The contradictions, however, are illusory, and the consequences are revolutionary.

Consider the first point. From the postulates one may infer that on the one hand light has the velocity c, and, on the other hand, even when, according to our traditional way of calculating, it should have the velocity $c + q$ (where q is the velocity of the source), its velocity is still c. Concretely, light from a source in motion with respect to a given frame of reference has the same velocity as light from a source at rest with respect to the same frame. (As one physicist suggested, this is as if we were to say that a man walking up an escalator does not get to the top any sooner than a man standing still on the escalator.) This seems absurd. But the reason it seems absurd is that we take it for grated that the velocity of the moving source must be added to the normal velocity of light to give the correct velocity of the beam emitted by the source. Suppose we abandon this assumption. We have already seen, after all, that motion has a queer effect on space and time measurements. It follow that the established notions of velocity must be reconsidered. The postulates were not inherently contradictory; the trouble lay with the classical laws of physics. They had to be changed. Einstein did not hesitate. To preserve his postulates he consigned the old system to the flames. In them were consumed the most cherished notions of space, time and matter.

One of the clichés about Einstein's theory is that it shows that everything is relative. The statement that every thing is relative is as meaningful as the statement that everything is bigger. As Russell pointed out, if everything were relative, there would be nothing for it to be relative to. The name "relativity" is misleading. Einstein was in fact concerned with finding something that is *not* relative, something that mathematicians call an invariant. With this as a fixed point, it might be possible to formulate physical laws that would incorporate the "objective residue" of an observer's experience; that is, that part of the space and time characteristics of a physical event which, though perceived by him, are independent of the observer and might therefore be expected to appear the same to all observers. The constancy principle of the velocity of light provided Einstein with the invariant he needed. it could be maintained, however, only at the expense of the traditional notion of time. And even this offering was not enough. Space and time are intertwined. They are part of the same reality. Tinkering with the measure of time unavoidably affects the measure of space.

Einstein, you will notice, arrived at the same conclusion as FitzGerald and Lorentz without adopting their electrical hypotheses. It was a consequence of his postulates that clocks and yardsticks yield different measurements in relative motion than at rest. Is this due to an actual physical change in the instruments? The question may be regarded as irrelevant. the physicist is concerned only with the difference in measurements. If clock springs and yardsticks contract, why is it not possible to detect the change? Because any scales used to measure it would suffer the same contraction. It has been said that this argument resembles the plan of the White Knight in *Through the Looking-Glass* "to dye one's whiskers green, and always use so large a fan that they could not be seen." But what is at issue is nothing less than the foundations of rational belief.

Science uses assumptions and theories to explain the physical world, and tests them by observation and experiment. When the evidence contradicts a theory, it is modified or discarded. For a long time it was useful and even comfortable to believe that length and time are absolute quantities. This was an assumption subject to verification. When the evidence failed to support the assumption, it had to be changed. Shrinking rulers and lagging clocks are a sad business; it affected men's minds like the Lisbon earthquake. The preacher had said that the earth would abide, but now no one could be sure. The subversion of space and time was no less shocking. but there was no alternative if physics was to remain a fruitful activity.

* * *

Earlier I mentioned Kaufmann's and Thomson's discovery that an electron increases in mass as it goes faster. Relativity explains this astonishing fact. The first postulate sets an upper limit to velocity. Specifically only light is mentioned, but one can deduce from the postulate that no material body can move faster than light. In Newton's system there were no such limits. He defined mass as the "quantity of matter," and gave the formula that acceleration (which means change—or, more accurately, rate of change—of velocity) is the quotient of the force acting on a body divided by its mass. In his view the mass of a body is the same at all speeds. But just as his laws of motion have been shown not to be universally true, his concept of the permanence of mass turns out to be flawed. According to Einstein's Special Theory, the resistance of a body to changes in velocity increases with velocity. Thus, for example, more force is required to increase a body's velocity from 50,000 to 50,001 miles per hour than from 100 to 101 miles per hour. The resistance to change is called *inertia*, and its measure is mass. (This jibes with the intuitive notion that the amount of force needed to accelerate a body depends on its "quantity of matter"—Newton's definition of mass.) The ideas fall neatly into place: with increased speed, inertia increases; increased inertia evinces itself as increased mass. The increase in inertia or mass is, to be sure, very small at ordinary speeds, and therefore undetectable, which explains why Newton and his successors, though a brilliant company, did not discover it. This circumstance also explains why Newton's laws are perfectly valid for all ordinary instances of matter in motion: even a rocket moving at 10,000 miles an hour is a tortoise compared to a beam of light at 186,000 miles a second. but the increase in mass becomes a major factor where high-speed nuclear particles are concerned; for example, the electrons in a hospital X-ray tube are speeded up to a point where their normal mass is doubled, and in an ordinary TV picture tube the electrons have 5 per cent extra mass due to their energy of motion. And at the speed of light the push of even an unlimited accelerating force against a body is completely frustrated, because its mass, in effect, becomes infinite.

It is only a step now to Einstein's fateful equation.

Energy is converted into mass by motion. The bookkeeping is quite exact. The quantity of additional mass, multiplied by an enormous number, the square of the speed of light, is equal to the energy that was turned into mass. In a moving body, therefore, mass and energy are equivalent. But is this equivalence of mass and energy a special circumstance attendant upon motion? What about a body at rest? does its mass also represent energy? Einstein concluded that it does. The Special Theory does not furnish a complete answer, but Einstein went beyond his theory. "The mass of a body is a measure of its energy content," he wrote in 1905, and gave his now famous formula, ($E = mc^2$, where E is energy content, m is mass (which varies according to speed) and c is the velocity of light.

"It is not impossible," Einstein said in this same paper, "that with bodies whose energy content is variable to a high degree (*e.g.*, with radium salts) the theory may be successfully put to the test." In the 1930s many physicists were making this test, measuring atomic masses and the energy of products of many nuclear reactions. All the results verified his idea. A distinguished physicist, Dr. E. U. Condon, tells a charming story of Einstein's reaction to this triumph: "One of my most vivid memories is of a seminar at Princton (1934) when a graduate student was reporting on researches of this kind and Einstein was in the audience. Einstein had been so preoccupied with other studies that he had not realized that such confirmation of his early theories had become an everyday affair in the physical laboratory. He grinned like a small boy and kept saying over and over, "*Ist dass wirklich so?*" Is it really true?—as more and more specific evidence of his $E = mc^2$ relation was being presented."

* * *

The year 1905 was for Einstein a marvellous year, like Newton's *annus mirabilis* (1665–6), when he discovered the law of gravitation, invented the calculus and laid the foundations of his great optical discoveries. Besides the paper on relativity, Einstein in 1905 wrote two other major papers: on the photoelectric effect, and on the dance of tiny particles in liquids and gases, a phenomenon known as Brownian motion. The paper on Brownians motion greatly illuminated the problem of molecular motion and provided a formula with which one could estimate the number of molecules in a given volume, though they themselves are invisible, by measuring the journeys of the tiny visible particles of liquid or gas. The other paper, for which he won the Nobel prize, brilliantly extended Max Planck's quantum theory by showing that light itself, like other forms of energy, consists of tiny packets (which later got the name photons).

For ten years after he formulated the Special Theory, Einstein grappled with the task of generalizing relativity to include accelerated motion. This essay cannot carry the weight of the details, but I shall describe the matter briefly.

While it is impossible to distinguish between rest and uniform motion by observations made within a system, it seems quite possible, under the same circumstances, to determine *changes* in velocity, i.e., acceleration. In a train moving smoothly in a straight line, at constant velocity, one feels no motion. But if the train speeds up, slows down or takes a curve, the change is felt immediately. One has to make an effort to keep from falling, to prevent the soup from sloshing out of the plate and so on. These effects are ascribed to what are called *inertial forces*, producing acceleration (the name is intended to convey the fact that the forces arise from the inertia of a mass, i.e., its resistance to changes in its state). It would seem, then, that any one of several simple experiments should furnish evidence of such acceleration, and distinguish it from uniform motion or rest. Moreover, it should even be possible to determine the effect of acceleration on a beam, of light. For example, if a beam were set parallel to the floor of a laboratory at rest or in uniform motion, and the laboratory were accelerated upward or downward, the light would no longer be parallel to the floor, and by measuring the deflection one could compute the acceleration.

When Einstein turned these points over in his mind, he perceived a loose end in the reasoning, which others had not noticed. How is it possible in either a mechanical or an optical experiment to distinguish between the effects of gravity, and of acceleration produced by inertial forces? Take the light-beam experiment. At one point the beam is parallel to the floor of the laboratory; then suddenly it is deflected. The observer ascribes the deflection to acceleration caused by inertial forces, but how can he be sure? He must make his determination entirely on the basis of what he sees *within* the laboratory, and he is therefore unable to tell whether inertial forces are at work—as in the moving train—or whether the observed effects are produced by a large (though unseen) gravitating mass.

Here then, Einstein realized, was the clue to the problem of generalizing relativity. As rest and uniform motion are indistinguishable, so are acceleration and the effects of gravitation. Neither mechanical nor optical experiments conducted within a laboratory can decide whether the system is accelerated, or in uniform motion and subjected to a gravitational field. (The poor wretch in tomorrow's space ship, suddenly thrown to the ground, will be unable to tell whether his vehicle is starting its rocket motor for the home journey or falling into the gravitational clutches of Arcturus.) Einstein formulated his conclusion in 1911 in his "principle of equivalence of gravitational forces and inertial forces."

His ideas invariably had startling consequences. From the principle of equivalence he deduced, among other things, that gravity must affect the path of a ray of light. This follows from the fact that acceleration would affect the ray, and gravity is indistinguishable from acceleration. Einstein predicted that this gravity effect would be noticeable in the deflection of the light from the fixed stars whose rays pass close to the huge mass of the sun. He realized, of course, that it would not be easy to observe the bending, because under ordinary conditions the sun's brilliant light washes out the light of the stars. But during a total eclipse the stars near the sun would be visible, and circumstances would be favorable to checking his prediction. "It would be extremely desirable," Einstein wrote in his paper enunciating the equivalence principle, "if astronomers would look into the problem presented here, even though the consideration developed above may appear insufficiently founded or even bizarre."

Eight years later, in 1919, a British eclipse expedition, headed by the famous astronomer Arthur Eddington, confirmed Einstein's astounding prediction. The weather on the small island of Principe was stormy, and the crucial photographs had to be taken through breaks in the clouds. For six nights Eddington and his companion developed the pictures, the days being spent in measurement. At last came a moment that Eddington later said was the greatest in his life: "The one plate I measured gave a result agreeing with Einstein."

Alfred North Whitehead described the occasion when the expedition results were announced to the Royal Society of London. "The whole atmosphere of tense interest was exactly that of the Greek drama: we were the chorus commenting on the decree of destiny as disclosed in the development of a supreme incident. There was dramatic quality in the very staging:—the traditional ceremonial, and in the background the picture of Newton to remind us that the greatest of scientific generalisations was now, after more than two centuries, to receive its first modification. Nor was the personal interest wanting: a great adventure in thought had at length come safe to shore."

* * *

Einstein saw with the eye of reason. He was so sure of himself that he did not feel he had to wait for the physical confirmation of his equivalence principle. In 1916 he announced his General Theory of Relativity, a higher synthesis incorporating both the Special Theory and the principle of equivalence. Two profound constructs are developed in the General Theory: the union of time and space into a four-dimensional continuum (a consequence of the Special Theory), and the curvature of space.

It was to one of his former professors at Zurich, the Russian-born mathematician Hermann Minkowski, that Einstein owed the idea of the union of space and time. "From henceforth," Minkowski had said in 1908, "space in itself and time in itself sink to mere shadows, and only a kind of union of the two preserves an independent existence." To the three familiar dimensions of space, a fourth, of time, had to be added, and thus a single new medium, space-time, replaced the orthodox frame of absolute space and absolute time. An event within this medium—one may, for example, think of a moving object as an "event"—is identified not only by three spatial co-ordinates denoting *where* it is, but by a time coordinate denoting *when* the event is there. *Where* and *when* are, as we have seen, judgments made by an observer, depending on certain interchanges of light signals. It is for this reason that the time co-ordinate includes as one of its elements the number for the velocity of light.

With the absolute space and time discarded, the old picture of the universe proceeding moment by moment from the past through the present into the future must also be discarded. In the new world of Minkowski and Einstein, there is neither absolute past nor absolute future; nor is there an absolute present dividing past from future and "stretching everywhere at the same moment through space." The motion of an object is represented by a line in space-time, called a "world line." The event makes its own history. The signals it emits take time to reach the observer; since he can record only what he sees, an event present for one observer may be past for another, future for a third. In Eddington's words, the absolute "here-now" of former beliefs has become a merely relative "seen-now."

But this must not be taken to mean that every observer can portray only his own world, and that in place of Newtonian order we have Einsteinian anarchy. Just as it was possible in the older sense to fix precisely the distance between two points in three-dimensional space, so it is possible in the four-dimensional continuum of space-time to define and measure distance between events. This distance is called an "interval" and has a "true absolute value," the same for all who measure it. Thus after all "we have found something firm in a shifting world."

How is the concept of curved space related to this picture? The concept itself sticks in the craw. A vase, a pretzel, a line can be curved. but how can empty space be curved? Once again we must think not in terms of metaphysical abstractions but of testable concepts.

Light rays in empty space move in straight lines. Yet in some circumstances (e.g., where the ray is close to the sun) the path of motion is seen to be curved. A choice of explanations offers itself. We may, for example, say that a gravitational mass in the neighborhood of the ray has bent it; or we may say that this gravitational mass has curved the space through which the ray is traveling. There is no logical reason to prefer one explanation to the other. Gravitational fields are no less an imaginative construct than space-time. The only relevant evidence comes from measuring the path of the light itself—not the field or space-time. It turns out to be more fruitful to explain the curved path of the light ray as an effect of curved space-time, rather than as an effect of the direct action of gravity on light.

Let me suggest an analogy. A thin sheet of rubber is stretched over a large drum kettle. I take a very light marble and permit it to roll over the sheet. I observe that the path of its motion is a straight line. I now take several lead weights and place them at different points on the rubber sheet. Their weight dimples it, forming small slopes and hollows. Suppose I release the marble on this surface. The path of motion will no longer be straight but will curve toward the slopes and eventually fall into one of the hollows. Now think of space-time as corresponding to the sheet of rubber, and large gravitational masses to the lead weights; think also of any "event"—a moving particle, a beam of light, a planet—as the counterpart of the marble rolling on the membrane. Where there are no masses, space-time is "flat" and paths of motion are straight lines. But in the neighborhood of large masses space-time is distorted into "slopes" and "hollows," which affect the path of any object entering upon them. This is what used to be called the attraction of gravitation. But gravitation in Einstein's theory is merely an aspect of space-time. The starlight bent toward the sun "dips" into the "slope" around it but has enough energy not to be trapped in the "hollow"; the earth circling the sun is riding on the "rim" of its "hollow" like a cyclist racing round a velodrome; a planet that gets too deep into the "hollow" may fall to the bottom. (This is one of the hypotheses astronomers make about collisions that may have formed new planets in our universe.) There are slopes and hollows wherever there is matter; and since astronomical evidence seems to favor the hypothesis that matter is on average uniformly distributed throughout the universe, and finite—though not necessarily constant—Einstein suggested the possibility that the whole of space-time is gently curved, finite, but unbounded. It is not inconsistent with this hypothesis that the universe is expanding, in which case the density of matter would decrease. A finite but unbounded universe is roughly analogous—though it is of higher dimension—to the two-dimensional curved surface of the earth. The area is finite without boundaries, and if one travels in a straight line in a given direction one must, after a time, return to the original point of departure.

* * *

Einstein spent his life searching for what is changeless in an incessantly changing world. He searched for unity in multiplicity. In his model of physical reality, space, time, energy, matter are bound together in a single continuum. The crown of his efforts—to find a set of field equations that would unite gravitational and electromagnetic phenomena—may have eluded him. But his achievement is beyond measure or praise. Two points about his work are worth making. The first is that his model of the world was not a machine with man outside it as observer and interpreter. The observer is part of the reality he observes; by observation he shapes it. "Event, signal and observation," as Bronowski has written, "is the relationship which Einstein saw as the fundamental unit in physics. Relativity is the understanding of the world not as events but as relations."

The second point is that his theory did much more than answer questions. As a living theory it forced new questions upon us. Einstein challenged unchallengeable writs; he would have been the last to claim that his own writs were beyond challenge. He broadened the human mind.

* * *

I want to say a few words about the man.

He was a thoroughly good man. He was kindly and tolerant. He was gentle. He felt for humanity and deeply sensed himself a part of it, sharing its hopes and fears. He hated war—the atavist instincts that produce it, the degradation it causes, the suffering it inflicts. By tragic chance one of his ideas paved the way for the most terrible weapon of all times. His innocent vision became a nightmare. For the last ten years of his life he abjured men to reflect upon their peril, to avert what he sadly suspected their folly would encompass—universal death.

He was a skeptical man, contemptuous of dogma, properly disrespectful of authority. His passion for understanding drove his researches. It was a passion not only to decode the patterns of the physical world, but also to grasp the deeper meaning of man's relation to nature and of man's dependence on man. He regarded science as a discoverer and a liberator. It would open the world and teach men how to live in it and to be happy.

He was a courageous man. He esteemed only truth and would not bow to tyrants, to mob opinion, to the clamor of irresponsible journalists, to the threats of squalid politicians. He respected the conscience of free-thinking men, whatever their beliefs, and he did not hesitate to speak out on behalf of heretics and others who in our unhappy time have been persecuted for their independence.

"It is the moral qualities of its leading personalities that are perhaps of even greater significance for a generation and for the course of history than purely intellectual accomplishments. Even these latter are, to a far greater degree than is commonly credited, dependent on the stature of character." Einstein's words, spoken on the occasion of a memorial celebration for Madame Curie, are appropriate to an estimate of his place in our civilization. The values he cherished are the values of science no less than of theirs. I do not mean one must be noble to be creative. I do mean that science as one of the glories of man is animated by the values of a decent society: the love of truth for its own sake, respect for dissent, independence, free expression. Einstein was an incorruptible human being. Scientific thought was for him the guide of action. He knew that, in Clifford's noble phrase, "the truth which it arrives at is not that which we can ideally contemplate without error, but that which we may act upon without fear." If the values he lived by were better served, mankind could look forward to a brighter future.

From James R. Newman, *Science and Sensibility* (New York: Simon & Schuster, 1961).

Questions and Topics for Discussion and Writing

1. What does Newman mean when he says that young scientists may have a flash of insight when they are young and spend the rest of their lives interpreting and elaborating their vision? Give some examples of this phenomena.

2. Why did scientists postulate the existence of the ether? What was it supposed to do?
3. What did the Michelson-Morley experiment show?
4. Newman says that Einstein took the philosophical position that nothing had meaning that could not be verified. Explain what this meant for Einstein's work.
5. Why is "relativity" a misleading name for relativity theory? What are the invariants in relativity theory?
6. How is the space-time continuum of relativity different from space and time in the Newtonian model of the world?
7. Newman says that Einstein was "properly disrespectful of authority." In what way was this disrespect essential to his work?

Charles Darwin

Issue

The Development of the Life Sciences

The life sciences have a long history, beginning with Aristotle's *History of Animals*, if not before. Parallel to the early development of biology is the development of medicine, which also has roots in the ancient Greeks and, of course, in many other cultures also. However, none of the early history of the life sciences will be considered in this book. Instead, we will jump right to the mid-nineteenth century and examine Darwin and the theory of evolution.

A strong argument can be made that biology really began to be a "scientific" discipline with Darwin, that the previous work in the biological subjects was interesting study but should not be called a science. That point need not be resolved here, but it certainly is true that the theory of evolution is one of the most interesting issues in the history of the life sciences from the point of view of its interaction with society.

In the *physical* sciences, the Copernican Revolution caused a considerable stir outside of scientific and scholarly circles because it called into question a matter of everyday life that had been taken for granted, that the earth was stationary and at the center of the universe. Most of the fuss about the Copernican viewpoint was not over its scientific credibility, but over the displacement of humanity from center stage. Likewise, in the *life* sciences, Darwin's theory took away yet another special place for humans and made us just another animal. In both cases objections to the theory were often made on religious grounds because the Bible, read literally, says that the sun moves (not the earth) and that God specifically created man and woman. The heliocentric theory of Copernicus has triumphed completely. Its present detractors are few and far between. Not so with evolution, which still has lively opponents who claim that evolution is not proved and that creationism is a viable alternate view.

Most of the articles in this section are concerned with the theory of evolution, its proponents and detractors. The final selection is an article on the beginnings of genetics. But these are far from being all of the articles in this book on the life sciences, as much of Unit III on the Environment and parts of Unit IV on Current Issues are about current problems in society that concern the sciences of living things.

In this section, we begin with "What Darwin Explained," by Byron Wall, which examines Darwin's work, what he tried to do and what he succeeded in doing. This is followed by one of the original and clearest statements of the theory of evolution by Darwin's rival and colleague, Alfred Russel Wallace. (Darwin himself is too long-winded.) Next is one of the most famous challenges to the teaching of evolution, the "Scopes Trial" in the State of Tennessee in

1925; the selection is an excerpt of the trial transcript. The theory of evolution and the evidence of evolution have both come a long way since the work of Darwin himself. The article "Darwinism Defined: The Difference between Fact and Theory" by prominent evolutionist Stephen Jay Gould brings the scientific issues up to date. The final essay, Ruth Moore's "The Discovery of Units of Heredity," reviews the work of Mendel and the early steps of a science of genetics. Without a viable theory of inheritance, Darwin was unable to explain the actual mechanism of evolution. Mendel's work, though done and published in Darwin's own time, was unknown to Darwin and his followers.

What Darwin Explained

Byron E. Wall

Darwin was a remarkable character. In his youth he travelled all around the world to exotic places, but then returned to England and lived for the rest of his long life as a recluse on his country estate. Early in his life he conceived the essential structure of the theory for which he is famous, but refrained from publishing it for years until his hand was forced by circumstance. Once his theory was published it caused an enormous stir among scientists and laymen, and in a short period of time came to dominate biological thought.

Given Darwin's eccentricities, the success of his approach warrants some examination. Just what did Darwin accomplish? What did the theory of evolution as he presented it explain?

To clarify these questions, first consider claims which might be made for Darwin but which are false.

Evolution Before Darwin

First, consider whether Darwin was an originator of ideas. Darwin was certainly not the first to consider that species had evolved. That idea had been around for ages. The most popular evolutionary theory current in Darwin's youth was that of Jean Lamarck, which Darwin had studied in detail. But there were many minor evolutionists around too, including Darwin's grandfather, Erasmus Darwin, all of whose ideas were known to Darwin.

But of course, Darwin's theory of evolution differed from these rivals by attributing the direction of evolutionary change to natural selection. Where others had claimed an inner perfectability or striving within living things to be the cause of modifications of structure, or attributed them to direct climactic or other environmental influences, Darwin saw the primary factor to be the weeding out of the unfit from the fit in populations of species that naturally are born with a wide variety of characteristics and capabilities. The unfit would simply not survive and propagate. But Darwin was hardly the first to think of nature operating this way.

Here for example is a well-known passage from Aristotle's *Physics* in which Aristotle discusses the distinction between necessity and accident in his analysis of purpose in nature. He sketches out the principle of natural selection, which he held to be erroneous, in order to show how some possible objections might be disposed of. The discussion of natural selection goes as follows:

> Why should nature work, not for the sake of something, nor because it is better so, but just as the sky rains, not in order to make the corn grow, but of necessity? What is drawn up must cool, and what has been cooled must become water and descend, the result of this being that the corn grows. Similarly if a man's crop is spoiled on the threshing-floor, the rain did not fall for the sake of this—in order that the crop might be spoiled—but that result just followed. Why then should it not be the same with the parts in nature, e.g. that our teeth should come up *of necessity*—the front teeth sharp, fitted for tearing, the molars broad and useful for grinding down food—since they did not arise for this end, but it was merely a coincident result; and so with all other parts in which we suppose that there is purpose? Wherever then all the parts came about just what they would have been if they had come to be for an end, such things survived, being organized spontaneously in a fitting way; whereas those which grew otherwise perished and continued to perish, as Empedocles says his "man-faced ox-progeny" did.[1]

[1] Aristotle, *Physics* Bk II Ch 8, 198b 17-33, in *The Basic Works of Aristotle*, ed. Richard McKeon, 249 (New York: Random House, 1941).

As a principle in and of itself, natural selection is clearly stated in this passage from Aristotle. It even appears to have been part of a full theory of evolution in the works of the Greek philosopher Empedocles, who lived about 450 B.C. Empedocles' own works have not survived intact. Scholars have pieced together fragments of his writings from various sources; they believe they have about one fifth of the original long poem that was the source of Aristotle's remarks. From those fragments it is obvious that the rest of Empedocles' evolutionary theory was very different from Darwin's. Empedocles speaks of disjointed limbs wandering around seeking other limbs to unite with and form whole bodies. Natural selection operated on the monsters that were formed from these chance unions of straying body parts. Small wonder that Aristotle had no trouble dismissing the idea.

More recent statements of natural selection also can be found. Darwin claimed to have hit upon the concept after having read Thomas Malthus' *Essay on Population*, but the concept is clearly expressed in Lyell's *Principles of Geology* and attributed to the French botanist Augustin de Candolle. In fact, in the *Origin of Species*, Darwin himself refers the idea to both Lyell and de Candolle.

Darwin on Heredity

What about heredity? If Darwin's theory of evolution explains how species change little by little over millions of years, it must also explain how they in general remain the same from generation to generation. Well, Darwin did have something to say on the subject of heredity, but it was not very profound and it was soon rejected. He called his idea the "hypothesis of pangenesis." In Darwin's own words, the basic idea was as follows:

> According to this hypothesis, every unit or cell of the body throws off gemmules or undeveloped atoms, which are transmitted to the offspring of both sexes, and are multiplied by self-division. They may remain undeveloped during the early years of life or during successive generations; and their development into units or cells, like those from which they were derived, depends on their affinity for, and union with other units or cells previously developed in the due order of growth.[1]

Actually even this poor hypothesis was more than Darwin said on the subject in the *Origin of Species*. In the first editions of the *Origin*, Darwin implicitly adopted the then current notions of blending inheritance, wherein the characteristics of both parents were somehow mixed together like different colours of ink in the offspring. A few years later it was pointed out that on this way of looking at things it was virtually impossible for natural selection to have any measurable effect on the characteristics of species. This so troubled Darwin that in later editions of the *Origin*, he softened his insistence on the importance of natural selection.

Apart from blending inheritance and pangenesis, Darwin had no way to explain why acorns don't grow into monkeys. Genetics as we know it is essentially a twentieth century science. With modern genetics, we can explain Darwin's theory of evolution, but the reverse in not true.

What about the origin of life itself? Despite the title, *On the Origin of Species*, Darwin had little to say on the subject. Darwin's choice of title refers to his account of how new species arise from earlier species. But for the original species, the first forms of life itself, Darwin believed that one or a few life forms were created directly by God. The concept of life arising directly from inorganic matter by natural causes is an extension of Darwin's theory beyond the limits he put on it himself.

The foregoing remarks on what Darwin did not accomplish were designed to clear the stage for the analysis of what he did accomplish and what his theory did explain. Before turning directly to that topic, we must digress once more to discuss the scientific and intellectual background of Darwin's work.

Darwin's Scientific Milieu

Darwin lived after the Copernican revolution. That fact has some consequences bearing on scientific explanations. In the article in this volume entitled "What the Copernican Revolution is All About," we looked at the dominant scientific world

[1] Charles Darwin, *The Descent of Man and Selection in Relation to Sex*, Ch VIII "Principles of Sexual Selection," Section: "The Laws of Inheritance." (London, 1871).

view before the Copernican revolution, in particular about Aristotle's separation of the universe into two parts, that below the moon and that beyond the moon. The sublunar world was characterized by change, generation and corruption, potentiality and actuality. All things had a beginning and an end, were born and died, and did so in fulfillment of some purpose. In the superlunar world, everything was eternal. The stars and planets moved endlessly in repeating cycles. Whatever purpose they had was fulfilled by maintaining their motion in a changeless pattern without determinate goals.

After the Copernican revolution, the distinction between above and below the moon disappeared. Galileo showed that the superlunar world displayed characteristics of the sublunar world, and Newton developed a unified physics that applied both to things on earth and to celestial bodies. But did this mean that the sublunar categories of fulfillment of purposes were then applied to the rest of the universe? In fact it was just the reverse; Aristotle's characterization of the celestial motions as continuing without final goals became the model for explanations both in the heavens and on earth.

Where Newton Saw Purpose at Work

Newton's system was based on a few simple axioms which he claimed to be universally true for all time and places and from which he could deduce logically necessary conclusions. For example, the principle of inertia stated that if a body was moving at a constant speed it would retain that speed in the same direction until some external force stopped it or altered its motion. Thus for both Newton and Aristotle, the planets moved continuously and eternally because it was logically necessary that they should do so, and not because they were seeking some goal. Of course, for Newton, their motion due to inertia would be in a straight line which had to be corrected by gravitational attraction to other bodies, such as the sun. But unlike Aristotle's conception, Newton's system applied with the same changeless, purposeless necessity to motions on earth.

There were certain aspects of events in the world that did not appear to Newton to follow necessarily from his premises and which therefore suggested that purposeful action was taking place. One was the principle of universal gravitation itself. Newton had no way of adequately explaining it in mechanical terms. He left open the question whether it was due to a material or an immaterial cause. That question was not given an answer until Einstein formulated general relativity theory in the twentieth century. Meanwhile it was largely forgotten that this *was* a problem. Newton cited two other indications of purpose afoot and claimed they showed evidence of Divine intervention. These were much noted and discussed after Newton's death. One concerned the solar system. Newton saw no necessity according to his physics that all the planets went around the sun in the same direction and all in approximately the same plane. Moreover, by his calculations, the planets ought not to be moving precisely as they had been observed over centuries to move, so something must be correcting their motions to keep the system going as it does. The other evidence of purpose was the living world. These two problem areas are discussed by Newton in this passage from the famous Query 31 of the *Opticks*:

> All material Things seem to have been composed of the hard and solid Particles...variously associated in the first Creation by the Counsel of an intelligent Agent. For it became him who created them to set them in order. And if he did so, it's unphilosophical to seek for any other Origin of the World, or to pretend that it might arise out of a Chaos by the mere Laws of Nature; though being once form'd it may continue by those Laws for many Ages. For while Comets move in very excentric Orbs in all manner of Positions, blind Fate could not make all the Planets move one and the same way in Orbs concentrick, some inconsiderable Irregularities excepted, which may have risen from the mutual Actions of Comets and Planets upon one another, and which will be apt to increase, till this System wants a Reformation. Such a wonderful Uniformity in the Planetary System must be allowed the Effect of Choice. And so must the Uniformity in the Bodies of Animals, they having generally a right and a left side shaped alike, and on either side of their Bodies two Legs behind, and upon their Shoulders, and between their Shoulders a Neck running down into a Back-Bone, and a Head upon it; and in the Head two Ears, two Eyes, a Nose, a Mouth, and a Tongue, alike situated. Also the first Contrivance of those very artificial Parts of Animals, the Eyes, Ears, Brain, Muscles, Heart, Lungs, Midriff, Glands, Larynx, Hands, Wings, swimming Bladders, natural Spectacles, and other Organs of Sense and Motion; and the Instinct of

> Brutes and Insects, can be the effect of nothing else than the Wisdom and Skill of a powerful everliving Agent, who being in all Places, is more able by his Will to move the Bodies within his boundless uniform Sensorium, and thereby to form and reform the Parts of the Universe, than we are by our Will to move the Parts of our own Bodies.[1]

Newton's argument is that since for both the planetary orbits and animal adaptation, events could have, under his principles of motion, turned out any number of ways, and since in fact they have turned out to be so well ordered, the particular way things are *must* have been chosen by some supreme intelligence. There is a curious leap in logic here. On the one hand, implicit in this argument is the assumption that the axioms of Newton's physics were merely *discovered* by Newton. The were *formulated* by God; or another way of putting it is that they are absolute principles of nature, not just the best way of thinking about it that a frail human mind could devise. That assumption is not unexpected. What may be more difficult for the modern reader is why Newton concludes that just because the way things are cannot be deduced as necessary consequences of nature's laws it must be that the way things are was chosen for some end by an agent able to do the choosing.

The reasoning seems to lie in Aristotle's account of causation. The passage quoted earlier from Aristotle having to do with natural selection was, as mentioned, part of a discussion of necessity and accident. He put forth the natural selection argument in order to dispose of it, but how did he dispose of it? His refutation is that the bodies and parts of animals are formed normally in a given regular fashion, but no things that happen spontaneously or by coincidence happen necessarily, and are therefore the necessary result of some cause. Aristotle recognized four categories of causes. By showing that animal adaptation did not follow necessarily from material, formal, and efficient causes, and also could not happen by chance, it must be that it follows necessarily from the remaining cause, the final cause or purpose. Newton's reasoning seems to be along the same lines.

The Design Argument

Darwin's age was completely dominated by the Newtonian viewpoint in science, especially in Great Britain. Newtonian physics was taken as revealed truth, at least in its general formulation, and along with it, Newton's analysis of what he called the effect of choice. Any regularity that could not be accounted for by necessary implication of physical laws was evidence of the hand of God.

This is the Design argument. The design in nature implied the existence of a designer. The study of nature, therefore, was analogous to the study of Scripture. Thus the study of nature could be, and was, called Natural Theology.

In 1829, the eighth Earl of Bridgewater died, leaving in his will a bequest for works "on the power, wisdom and goodness of God as manifested in the Creation." In other words, for works of Natural Theology that developed the Design argument. Eight "Bridgewater Treatises" were written by notable authorities in the 1830s. One of the most famous was by Sir Charles Bell, a distinguished anatomist at the University of Edinburgh. It bore the title, *The Hand: Its Mechanism and Vital Endowments as Evincing Design.* Bell's approach was to consider not just the human hand, but to compare the structure and function of many animals that had hand-like appendages, and to show that every difference in visible structure was accompanied by a multitude of internal relations that made the whole animal adapted to its function in nature. Hence only God could have designed it so. As Bell expressed it,

> It must now be apparent that nothing less than the Power which originally created is equal to effect those changes on animals, which adapt them to their conditions: and that their organization is predetermined; not consequent on the condition of the earth or of the surrounding elements...
>
> It has been shown, that...there is nothing like chance or irregularity in the composition of the system. In proportion indeed as we comprehend the principles of mechanics, or of hydrolics, as applicable to the animal machinery, we shall be satisfied of the perfection of the design. If anything appear disjointed or thrown in by chance, let the student

[1] Isaac Newton, *Opticks,* 4th ed., Bk III, Pt. I, Query 31 (London, 1704).

> mark that for contemplation and experiment, and most certainly, when it comes to be understood, other parts will receive the illumination, and the whole design stand more fully disclosed.[1]

The force of the design argument according to its own criteria depended upon the impossibility of accounting for the evident design with uniform physical laws. But by Darwin's age, many of the mysteries of nature *had* been deduced from specifiable physical laws, even some of those mysteries mentioned by Newton.

The Nebular Hypothesis and Uniformitarianism

Newton's conclusion that the planets would not stay precisely in their orbits due to gravitational effects alone was shown to be false by the French mathematical astronomer, Pierre Laplace. Laplace carried the calculations further and showed that while the planets do deviate from their mean orbits, the combined gravitational effect of the different bodies in the system bring them back to their regular paths over time. In answer to Newton's other planetary mystery, Laplace offered a solution which came to be called the nebular hypothesis. The reason that all the planets move in the same direction, lie in about the same plane, and indeed rotate in the same direction on their individual axes, Laplace suggested, was because the solar system was originally a mass of hot gas which rotated about some central axis. As the gas cooled and condensed, it formed itself into the sun, planets and their satellites, which, due to the conservation of angular momentum, continued to rotate in the same direction. As far as Laplace was concerned, he no longer needed the hypothesis of God.

Closer to home, another realm in which design was being accounted for by physical law was geology. The Uniformitarian account of the shape of the earth, set forth in magnificent detail in Charles Lyell's *Principles of Geology*, which Darwin read aboard H.M.S. Beagle, argued that mountains, valleys, rivers, seas, deserts, forests, and so on, were not created as they are now, but instead formed naturally over eons of time from such causes as wind, rain, erosion, earthquakes, droughts, etc.—in short, by the slow accumulation of small changes, all wrought by motions that do follow uniform universal laws.

Uniformitarianism was but one of the new geological theories that began to appear in the late eighteenth century, partly as the result of new knowledge of the earth's shape and structure that followed from voyages of discovery to distant lands and from the churning up of land at home in the transformation of the environment during the industrial revolution. The new information did not confine itself to rock formations. It also provided a rich source for a new problem of interpretation through the discovery of fossil remains of presumably extinct species. The kinds of fossils found differed from layer to layer in the earth's strata. In the deeper strata were found fossils of animals no longer known. In the less deep, the fossils resembled living species more, and in the shallowest strata they resembled living species most. This pattern was especially marked in the remains of the higher vertebrates—those closest to the human anatomy.

A uniformitarian way of coping with the fossil evidence was to deny it. Lyell himself argued, unconvincingly, that what appeared to be new species were really the result of migrations of species that previously existed in other parts of the world. All the species were created at the beginning. The theistic aspect of this interpretation was that the Creator worked his wonders through natural laws, and did not have to intervene in the system. But for those not wedded to uniformitarianism, the evidence of new species was further evidence of God at work. Species were created, not all at once, but in successive stages as conditions became fitting for them to exist. This view could even be accommodated to the creation story in the Book of Genesis by taking the seven days of creation to be metaphors for different epochs of the earth's history.

Darwin's Dilemma

Now, taking the historical sketch as a general context, let us turn to Darwin himself. The events of Darwin's early career posed a dilemma to him. Consider the following hypothetical question.

[1] Charles Bell, *The Hand: Its Mechanism and Vital Endowments as Evincing Design* (London, 1833).

Suppose yourself to be a young English chap named Charles Darwin. You go to Cambridge in the early nineteenth century and while you are there you learn a lot about animal and plant adaptation, about the design argument and the evidence for it, under the benevolent tutelage of a clergyman-naturalist. You learn above all else to examine the evidence of nature slowly and carefully before leaping to conclusions. You take a position as a naturalist on a five-year voyage around the world, during which you see with your own eyes more and more of the diversity of nature and the amazing balance of flora and fauna in each region you visit.

You especially notice the propensity of species to exhibit variations in their external characteristics, and you also notice that those variations differ from region to region. You know that all this can be accounted for by supposing special creations of species in different times and places. You also know of theories that account for species differences due to the effects of the surrounding environment. But on some islands, miles apart, you see the differences in the animal and plant life occurring between islands that have much the same environmental conditions.

You have taken with you Lyell's *Principles of Geology*, and part of your duties are to do geological surveys. Your work makes a competent geologist of you, and you see a great deal of evidence to support the uniformitarian view that you read expounded in Lyell's book. But you also learn a lot about fossils of extinct species and note their relationship to living species in the same areas. You suspect that the fossils represent an earlier form that has become modified over time into the present living species.

Eventually you return home, write up your observations as required, and begin to ponder what you have seen and read. The inorganic world *does* seem to be governed by uniform laws. The living world *does* exhibit a truly amazing ordered design. If species have been modified over time so as to be wonderfully adapted to their environments, perhaps the cause of their modifications is the same set of physical laws that govern geological changes. But what about those islands? Uniform laws are supposed to work uniform effects.

Suppose after mulling this over for a couple of years, you happen to read a certain essay on population by Thomas Malthus and that gives you an idea, or reminds you of one you read somewhere else and have forgotten. You conceive the principle of natural selection. You see that if each generation of living things varies naturally, some of the variations will be more favourably suited to surviving in the *physical* environment than others, and also, some variations will be more favourably suited for surviving in the competitive *living* environment. The cumulative effect over several generations can explain why species appear to be adapted to their physical surroundings, and it can also explain the differences on those isolated islands having the same physical conditions, because the living environment was different on each island—different predators, different sources of food, different proportions of each species. The uniform cause of the evolution of species is neither the physical nor the living environment, but the variability of the species themselves. Natural selection is merely a result of conditions, and it merely happens to direct the course of evolution the way it does. There's nothing supernatural about it.

Now, suppose all this is clear to you in 1838. Would you do as Alfred Russell Wallace did when he came to the same idea twenty years later and rush to announce it to the world? Or would you do as Charles Darwin did when he reached this position and sit on it for twenty years? Perhaps if you had been through what Darwin had been through and you knew what Darwin knew, you would not rush your idea into print.

Many reasons have been offered to explain why Darwin kept his idea to himself, reasons having to do with Darwin's personality, his childhood, his overbearing father, his timidity about open controversy, his wish not to offend his wife by the theological implications that could be drawn from his theory, and so on. These are valuable insights, but put them aside and consider the case purely on its own merits.

Darwin's Failure to Explain Evolution

Natural selection *could* explain the species phenomena that Darwin had observed. But it *did not* explain them. Darwin knew enough to know that it did not. There were too many other possible explanations. Consider the evidence itself. Darwin had seen a lot and read a lot, but of course he had not encountered the whole living world. The fossil record was woefully incomplete. Maybe all species were created at once, and just appeared in different places at different times. Even if they had originally appeared at different times and places, they could have been specially created then and there. Even if natural selection could explain the preservation of some favourable characteristics, it did not necessarily explain all of the amazing adaptive design exhibited by living forms.

Then there is variation in species. Darwin had seen that the offspring of species do exhibit varying characteristics, but he did not know what the extent of this variation was, nor did he have any idea of the cause of variation. Could it continue indefinitely or was it restricted to a limited range?

Laplace had been able to show that the apparent design in the solar system was the necessary result of uniform laws because he could *calculate* the results from those laws. (Actually Laplace did not quite succeed to the extent he thought he did.) But Darwin neither knew the relevant laws nor had any means to calculate their effects. The process of induction is quite different in the living as opposed to the inorganic world. In the latter, it is not unreasonable to assume that the properties exhibited by a few samples of a substance will also be exhibited by all of that substance. But in the living world, every instance is somewhat different, even within the same species. A pound of lead is a pound of lead, but every rose is unique—except in poetry.

What, then, would be accomplished by tossing out one more speculative theory of evolution to the world? It proved nothing. To prove the theory Darwin would have to accomplish two tasks: On the one hand, he would have to be able to state the detailed physical and biological laws that are the cause of variation. He would need a detailed knowledge of reproduction and heredity. He would need to be able to show exactly how natural selection works, to specify how long is required for successive stages of evolution and the factors which determine the direction of evolution. On the other hand, he would have to show that the results he could deduce did indeed happen. He would have to reconstruct the natural history of the world; show what species existed when and where, when they became extinct, when new species arose, what their different characteristics were. For this he would have to rely on a multitude of indirect evidence to interpret the fossil record and fill in the gaps. He would need many subsidiary theories to establish the validity of the historical record that was put together on this indirect evidence.

Darwin completed neither of these tasks. They were far beyond the powers of any one man. That is what was required to prove the theory of evolution, and Darwin failed to do it. But that is precisely what he was trying to do!

With immense patience, Darwin sat down to his tasks and worked away at them piecemeal, as though he would live forever. He spent years dissecting and classifying barnacles. The patient matter-of-fact way he went about his study of barnacles is revealed in the anecdote told about one of his young children during these years. The child took it for granted that dissecting barnacles was a perfectly ordinary and normal thing for a father to do. Upon visiting the home of a playmate and being shown around the living quarters, the child detected that something was missing and exclaimed, "But where does your father do *his* barnacles? " But barnacles were only one of many studies undertaken by Darwin. He did extensive experiments cultivating orchids and breeding pigeons to study the variation process. He had long discussions with animal breeders to learn more about the rules that govern the preservation of desirable characteristics. He read widely in geology, paleontology, zoology and botany—and not just what were considered scientific treatments of these subjects; he also consulted such publications as gardening magazines to gain practical knowledge that others might not think important. He kept up a correspondence with leading scientists to pick their brains.

He published some of his subsidiary studies and gained a modest reputation, but except for a few trusted colleagues he kept his main theoretical conception to himself. Twenty years had elapsed since Darwin read Malthus and got his flash of insight, and he was still at work collecting evidence and working out the details of his theory. Twice he had written up a tentative sketch of his theory of evolution, but he kept these short essays to himself. At the urging of the colleagues who knew of his theory, he had reluctantly begun to write out the details when disaster struck.

Wallace Rushes In

The young upstart naturalist Alfred Russel Wallace, whom Darwin had previously corresponded with, sent Darwin a short paper that Wallace had hurriedly composed on an idea that occurred to him after reading Malthus' *Essay on Population.* The essay was titled "On the Tendency of Varieties to Depart Indefinitely from the Original Type." As Darwin remarked at the time, it was an excellent short abstract of one of Darwin's brief sketches of his own theory, which of course Wallace knew nothing about.

With the threat of Wallace completely upstaging him and getting all the credit, Darwin finally took the leap and published his own major theory and research. At first, an excerpt from the second of his sketches was published along with Wallace's paper by the Linnean Society. Then, the next year, 1859, Darwin published a longer, more detailed version of his theory. It was not the huge, multivolume treatise that Darwin had begun to write, to be entitled "Natural Selection." Instead, because of that blasted Wallace, he had to make do with a mere 490 pages, which he wished to give the cautious title of "An Abstract of an Essay on the Origin of Species." Pushed to a little more boldness, the full title of the book as it first appeared was *On the Origin of Species by Means of Natural Selection; or, the Preservation of Favoured Races in the Struggle for Life*. In the sixth edition of the book, the word "On" was dropped from the main title, which gives the impression that it is intended as a definitive treatment of the subject. The book was a great success and made Darwin famous. By the time of the sixth edition, Darwin probably did feel it was definitive.

It is important to remember, however, that Darwin never did complete what he set out to do. He did publish other books that would have been individual volumes of his planned great work and he revised the *Origin* five times, but with all of this he fell short of his original aim—a logically flawless, deductive theory of evolution by natural selection.

If Darwin failed to achieve his great goal, it is a fair question to ask just what he did achieve. The theory of evolution is now regarded as correct and plays a central role in modern biology, though many of the details of Darwin's own account and even some of his major conceptions have been discarded, particularly Darwin's view of inheritance.

What Darwin Did Explain

Darwin certainly deserves the credit for stating his theory in enough detail to make people take him seriously, and to prompt the further research that has led to the establishment of evolutionary theory. Darwin had to be taken seriously because of the sheer amount of evidence he amassed that favoured his interpretation. Philosophers have frequently remarked that the theory of evolution has become established not by infallible deductive logic, but because the evidence for it far outweighs the evidence for any other interpretation. Historians have interpreted Darwin's long delay in publishing the theory to a desire to overwhelm the opposition. Certainly Darwin showed himself to be thinking along these lines. In the introduction to the *Origin of Species*, Darwin, apologizing to the reader for the omission of many relevant details, expressed his hope of someday publishing all the facts and references on which he based his conclusions. "For," he stated, "I am well aware that scarcely a single point is discussed in this volume on which facts cannot be adduced, often apparently leading to conclusions directly opposite to those at which I have arrived. A fair result can be obtained only by fully stating and balancing the facts and arguments on both sides of each question..."[1]

So, on this way of looking at it, Darwin's accomplishment was to tip the scales of evidence in favour of evolution, and in so doing spawn so much research that confirmed his theory that other views were driven to extinction. This might be called natural selection among scientific theories, or the survival of the fittest *theory*. But here is another accomplishment of Darwin, one which he himself may have been unaware of. It concerns a point of logic.

Darwin's Contribution to Logic

The opposition to Darwin's theory can be divided into three main types: emotional, scientific and logical. The emotional opposition stemmed from

[1] Charles Darwin, *On the Origin of Species By Means of Natural Selection,* 1st ed., 1-2 (London, 1859).

repugnance to the idea that man may have descended from some lower animal form. On the subject of *human* evolution, Darwin wrote to Wallace that though it was "the highest and most interesting problem for the naturalist," it was "so surrounded with prejudices" that he thought he would avoid it altogether. He finally did go ahead and write *The Descent of Man*. Nevertheless, Darwin let others fight the emotional furor that followed. He had no interest in any other than scientific disputes. The scientific opposition stemmed from other interpretations of the phenomena of nature. This is the arena in which Darwin's weight of evidence counted most, where he produced so many facts that could be interpreted by *his* theory, but not by other evolutionary or non-evolutionary theories.

The third type of opposition, the logical, lurked behind the other two. It has to do with the design argument. The design argument gave solace to those who would oppose Darwin on human evolution because it implied that God had specially intervened in his system to create all species, including man. The same argument propped up rival scientific theories because it held that the complexities of animal and plant adaptation could not have arisen from natural causes.

Consider once again of the logic of the design argument. It does *not* assert that it is highly improbable that complex living things were formed solely according to simple physical laws. It asserts that it is *impossible* for them to have done so. It is impossible because it would require many coincidences, and things that happen as a result of coincidence or chance do not happen with regularity. Since living things exhibit marvelous regularity in their structures and fit so well together in the order of nature, they could not have resulted from coincidental events that took place for no purpose and were merely the result of random physical motion.

No amount of evidence can assail this argument until the fallacy in it is seen. For me, Darwin's greatest achievement was to reveal that fallacy. The fallacy is that the design argument fails to take account of time. Random causes, such as whatever caused the variations in plants and animals that Darwin observed, produce random effects—in the *first* instance. But if one of those random variations, however infrequent, has a greater tendency to be preserved than other variations, then in successive repetitions of the generating process that variation becomes more and more frequent until its occurrence becomes the regular, expected effect.

In Sir Charles Bell's "Bridgewater Treatise" on *The Hand*, the staggering complexity and intricacies of structure of hand-like appendages in different animals that make them so perfectly adapted to their functions provide all the evidence of design necessary for Bell to conclude that only God could have made them so; they could certainly not have arisen as chance artifacts of natural processes. We can compare that argument with one given by Darwin in *The Origin of Species*, where the instrument under discussion is not the hand, but one of even greater complexity and perfection, the eye. In the passage below, notice that the thrust of Darwin's argument is not to prove that the eye evolved from natural causes, but only to defeat the logic of the design argument by showing that it is possible that it did evolve naturally.

Darwin begins as follows:

> To suppose that the eye, with all its inimitable contrivances for adjusting the focus to different distances, for admitting different amounts of light, and for the correction of spherical and chromatic aberration, could have been formed by natural selection, seems, I freely confess, absurd in the highest possible degree. Yet reason tells me, that if numerous gradations from a perfect and complex eye to one very imperfect and simple, each grade being useful to its possessor can be shown to exist; if further, they eye does vary ever so slightly, and the variations be inherited, which is certainly the case; and if any variation of modification in the organ be ever so useful to an animal under changing conditions of life, then the difficulty of believing that a perfect and complex eye could be formed by natural selection, though insuperable by our imagination, can hardly be considered real. How a nerve comes to be sensitive to light, hardly concerns us more than how life itself first originated; but I may remark that several facts makes me suspect that any sensitive nerve may be rendered sensitive to light...[1]

Darwin goes on to give some examples of light

[1] *Ibid.*, 186-87.

sensitive nerves in various crustaceans, showing that eye-like contrivances exist in many degrees of complexity and relative perfection. This is to establish the plausibility of the idea that something like an eye evolved from a mere light-sensitive nerve. Then Darwin continues:

> With these facts, here far too briefly and imperfectly given, which show that there is much graduated diversity in the eyes of living crustaceans, and bearing in mind how small the number of living animals is in proportion to those which have become extinct, I can see no very great difficulty...in believing that natural selection has converted the simple apparatus of an optic nerve merely coated with pigment and invested by transparent membrane, into an optical instrument as perfect as is possessed by any member of the great Articulate class.
>
> He who will go thus far, if he find on finishing this treatise that large bodies of facts, otherwise inexplicable, can be explained by the theory of descent, ought not to hesitate to go further, and to admit that a structure even as perfect as the eye of an eagle might be formed by natural selection, although in this case he does not know any of the transitional grades. His reason ought to conquer his imagination; though I have felt the difficulty far too keenly to be surprised at any degree of hesitation in extending the principle of natural selection to such startling lengths.[1]

In both Darwin's analysis of the eye and Bell's analysis of the hand, the chief source of evidence is comparative anatomy of eyes and hands of living creatures. Bell's argument is that this great complex diversity proves the existence of a purposeful agent of design. Darwin's argument is that it doesn't prove that, because it is *possible* that these marvelous effects are the result of gradual evolution by natural selection over enormous periods of time.

Time as a factor in a scientific theory was not new with Darwin. The uniformitarian geology that guided Darwin's thoughts depended on the accumulation of small causes to produce large effects. But while the arrangement of geological things in the world was noticeably different from ages past, the uniformitarians did not claim that the level of order in the present was different from that in the past. The nebular hypothesis required a vast amount of time in its explanation of how the apparent order of the solar system arose from a chaos of gas. But the order that was to be explained was the direction of orbit and rotation of the planets, and that was already present in the rotating nebula that was its former state. Similarly, the chief rival theory of evolution in Darwin's day held that species slowly changed over long periods of time, but the changes were effected by the individuals in each generation adapting themselves to the environment and then passing on that adaptation to the next generation. Hence the chief phenomenon to be explained, from the point of view of the design argument, the *orderedness* of living things, was given to them as a capacity from the start.

In Darwin's use of time, the resulting order is far greater than that of its causes, and the random causes are unrelated to the ordered effects. Darwin of course did not point out the fallacy in the design argument in this abstract fashion; he presented his case in terms of concrete biological phenomena. Because he argued in concrete terms, the truth or falsity of his conclusions depended on the facts. Because he did not have all the facts, he could not prove he was right, as he readily admitted. But his logic stands, independent of the theory of evolution.

The importance of the logical argument is that it not only cleared away the obstacle of the design argument from the advance of biological theory, it also opened new vistas for many subjects in which long sequences of events might produce unexpected results. The great popularity of Darwin's theory in the late nineteenth century assured that both his conclusions and his arguments would get careful scrutiny. From direct analysis of Darwin's works and from that indirect process through which bits and pieces of the thought of an influential person get assimilated into the common notions of an era, an evolutionary approach has become part of the analysis of many problems that have nothing to do with biological species.

An example is the work of Darwin's son, Sir George Darwin on lunar evolution. The younger Darwin argued that the reason that the moon is always turned the same way toward the earth, instead of rotating on its axis, is due to the long-term

[1] *Ibid.*, 188.

effect of the tides on earth. It is the moon which causes the tides in the first place, but since the earth rotates, the gravitational pull of the earth produces a slight drag on the moon. That drag, argued George Darwin, could suffice to slow down and eventually stop a rotation of the moon on its own axis. The subject matter was not biological, but the reasoning followed the pattern of his father's logic, namely to show that the known effect *can* follow the proposed cause, given sufficient time. Then anyone who wishes to oppose the suggested theory cannot do so by claiming it is impossible.

Conclusion

This chapter began by robbing Darwin of credit for either originating or proving the theory of evolution. Now it seems on the verge of giving him credit for half the important ideas of the last one hundred years. Let's stop short of that. Evolutionary thinking was commonplace throughout the nineteenth century. It is absurd to suggest that ever since 1859, everyone who thought of change over time got the idea from Darwin. What Darwin does deserve credit for is laying a solid groundwork for the later establishment of the theory of evolution by natural selection, causing the demise of the design argument, and turning the attention of scientists, social scientists, philosophers, and the general public to a closer look at the effect of time. To understand that, think about the overriding difference between the results predicted by Darwin's theory and by other evolutionary concepts. The other evolutionary theories, prominent in the nineteenth century, including not only biological theories but also social theories, had a common view that evolution meant change in a given direction, in short, progress. Their conceptions of the cause of evolution imply a certain direction due to an inner striving and the like.

What Darwin explained was that systematic change over time did not imply progress. The process Darwin saw was a blind production of diversities pruned down by a ruthless competition. The process may give the appearance of progress because it can result in increasing order within the system. But there is no inherent direction to the process and no inherent reason to suppose that natural changes are changes for the better. According to Darwin, when the environment changes, those who happen to have what it takes to survive do, and all the others just become extinct. Design and order in the world today is an accident of circumstance, not a product of forethought, and there is no guarantee that the order of today will be of much use in the world of tomorrow. That is the most important thought that Darwin explained to us.

Byron E. Wall

Questions and Topics for Discussion and Writing

1. What are some of the accomplishments and explanations often attributed to Darwin, but for which he was not responsible?
2. Explain the principle of natural selection.
3. What did Laplace mean when he said he no longer needed the hypothesis of God?
4. Define uniformitarianism. Why was the evolution of species a problem for uniformitarianism?
5. Discuss the design argument. How did it arise; what did it explain; what is wrong with it?
6. Discuss and compare the concepts of evolution and progress.
7. Why is time such a crucial component of the theory of evolution?
8. The author asserts that Darwin's greatest achievement was to establish a point in logic, not to establish the theory of evolution. Is he being serious? How could the point in logic possibly outweigh the enormous work that Darwin devoted his entire life to? Make a case for or against the author's viewpoint.

* On the Tendency of Varieties to Depart Indefinitely from the Original Type

Alfred Russel Wallace

In 1858, Charles Darwin had been patiently working up his theory of evolution by natural selection for twenty years without publishing anything on the subject—despite pleas by several of his scientific colleagues who knew of his work that he should commit some of it to print. Then a young naturalist acquaintance, Alfred Russel Wallace, who was then working in Indonesia, sent him a paper he had written, asking for Darwin's comments and asking Darwin to forward it on if he thought it worthy of publication. The paper, "On the Tendency of Varieties to Depart Indefinitely from the Original Type," contained to Darwin's amazement, a more or less complete statement of the theory of evolution by natural selection that Darwin had been labouring over all these years and was not yet ready to publish.

Faced with the possibility of being completely upstaged by his younger colleague, Darwin was finally persuaded to publish *something.* A few months later, on July 1, 1858, Wallace's paper was read to the Linnean Society (and then published in its Journal) and along with it, two pieces by Darwin, one an extract from his notebook of 1839, the other an abstract of a letter Darwin wrote to the American biologist, Professor Asa Gray, in 1857. The main purpose of publishing Darwin's pieces was to show that he had been thinking along the same lines as Wallace long before Wallace had. They certainly would not have been satisfactory to Darwin as statements of his theory.

The next year, 1859, Darwin, compelled now to commit himself, published what *he* viewed as merely an abstract of the work he was writing. The abstract, running 490 pages, is the work we know as *On the Origin of Species,* and is the work which marks the beginning of biology as a modern science. Clearly Darwin is not a man of few words. It is difficult to do justice to Darwin in a collection such as this by excerpting a few pages from one of his works. Those readers with more than a passing interest in the issues should certainly consult *The Origin of Species* itself, and one of Darwin's several other subsequent works, to see the power and depth of his arguments and explanations. Here, to serve the purpose of giving the reader a sense of the theory of evolution as it was understood by Darwin is the paper by Wallace as it appeared in the *Journal of the Linnean Society* in 1858. It was, after all, even in Darwin's view, an excellent précis of Darwin's own work.

One of the strongest arguments which have been adduced to prove the original and permanent distinctness of species is, that *varieties* produced in a state of domesticity are more or less unstable, and often have a tendency, if left to themselves, to return to the normal form of the parent species; and this instability is considered to be a distinctive peculiarity of all varieties, even of those occurring among wild animals in a state of nature, and to constitute a provision for preserving unchanged the originally created distinct species.

In the absence or scarcity of facts and observations as to *varieties* occurring among wild animals, this argument has had great weight with naturalists, and has led to a very general and somewhat prejudiced belief in the stability of species. Equally general, however, is the belief in what are called "permanent or true varieties,"—races of animals which continually propagate their like, but which differ so slightly (although constantly) from some other race, that the one is considered to be a *variety* of the other. Which is the *variety* and which the original *species*, there is generally no means of determining, except in those rare cases in which the one race has been known to produce an offspring unlike itself and resembling the other. This, however, would seem quite incompatible with the "permanent invariability of species," but the difficulty is overcome by assuming that such varieties have strict limits, and can never again vary further from the original type, although they may return to it, which, from the analogy of the domesticated animals, is considered to be highly probable, if not certainly proved.

It will be observed that this argument rests entirely on the assumption, that *varieties* occurring in a state of nature are in all respects analogous to or even identical with those of domestic animals, and are governed by the same laws as regards their permanence or further variation. But it is the object of the present paper to show that this assumption is altogether false, that there is a general principle in nature which will cause many *varieties* to survive the parent species, and to give rise to successive variations departing further and further from the original type, and which also produces, in domesticated animals, the tendency of varieties to return to the parent form.

The life of wild animals is a struggle for existence. The full exertion of all their faculties and all their energies is required to preserve their own existence and provide for that of their infant offspring. The possibility of procuring food during the least favourable seasons, and of escaping the attacks of their most dangerous enemies, are the primary conditions which determine the existence both of individuals and of entire species. These conditions will also determine the population of a species; and by a careful consideration of all the circumstances we may be enabled to comprehend, and in some degree to explain, what at first sight appears so inexplicable—the excessive abundance of some species, while others closely allied to them are very rare.

The general proportion that must obtain between certain groups of animals is readily seen. Large animals cannot be so abundant as small ones; the carnivora must be less numerous than the herbivora; eagles and lions can never be so plentiful as pigeons and antelopes; the wild asses of the Tartarion deserts cannot equal in numbers the horses of the more luxuriant prairies and pampas of America. The greater or less fecundity of an animal is often considered to be one of the chief causes of its abundance or scarcity; but a consideration of the facts will show us that it really has little or nothing to do with the matter. Even the least prolific of animals would increase rapidly if unchecked, whereas it is evident that the animal population of the globe must be stationary, or perhaps, through the influence of man, decreasing. Fluctuations there may be; but permanent increase, except in restricted localities, is almost impossible. For example, our own observation must convince us that birds do not go on increasing every year in a geometrical ratio, as they would do, were there not some powerful check to their natural increase. Very few birds produce less than two young ones each year, while many have six, eight, or ten; four will certainly be below the average; and if we suppose that each pair produce young only four times in their life, that will also be below the average, supposing them not to die either by violence or want of food. Yet at this rate how tremendous would be the increase in a few years from a single pair! A simple calculation will show that in fifteen years each pair of birds would have increased to nearly ten millions! whereas we have no reason to believe that the number of the birds of any country increases at all in fifteen or in one hundred and fifty years. With such powers of increase the population must have reached its limits, and have become stationary, in a very few years after the origin of each species. It is evident, therefore, that each year an immense number of birds must perish—as many in fact as are born; and as on the lowest calculation the progeny are each year twice as numerous as their parents, it follows that, whatever be the average number of individuals existing in any given country, *twice that number must perish annually*,—a striking result, but one which seems at least highly probable, and is perhaps under rather than over the truth. It would therefore appear that, as far as the continuance of the species and the keeping up the average number of individuals are concerned, large broods are superfluous. On the average all above *one* become food for hawks and kites, wild cats and weasels, or perish of cold and hunger as winter comes on. This is strikingly proved by the case of particular species; for we find that their abundance in individuals bears no relation whatever to their fertility in producing offspring. Perhaps the most remarkable instance of an immense bird population is that of the passenger pigeon of the United States, which lays only one, or at most two eggs, and is said to rear generally but one young one. Why is this bird so extraordinarily abundant,[1] while others producing two or three times as many young are much less plentiful? The explanation is not difficult. The food most congenial to this species, and on which it thrives best, is abundantly distributed over a very extensive region, offering such difference of soil and climate, that in one part or another of the area the supply never fails. The bird is capable of a very rapid and long-continued flight, so that it can pass without fatigue over the whole of the district it inhabits, and as soon as the supply of food begins to fail in one place is able to discover a fresh feeding-ground. This example strikingly shows us that the procuring a constant supply of wholesome food is almost the sole condition requisite for ensuring the rapid increase of a given species, since neither the limited fecundity, nor the unrestrained attacks of birds of prey and of man are here sufficient to check it. In no other birds are these peculiar circumstances so strikingly combined. Either their food is more liable to failure, or they have not sufficient power of wing to search for it over an extensive area, or during some season of the year it becomes very scarce, and less wholesome substitutes have to be found; and thus, though more fertile in offspring, they can never increase beyond the supply of food in the least favourable seasons. Many birds can only exist by migrating, when their food becomes scarce, to regions possessing a milder, or at least a different climate, though, as these migrating birds are seldom excessively abundant, it is evident that the countries they visit are still deficient in a constant and abundant supply of wholesome food. Those whose organization does not permit them to migrate when their food becomes periodically scarce, can never attain a large population. This is probably the reason why woodpeckers are scarce with us, while in the tropics they are among the most abundant of solitary birds. Thus the house sparrow is more abundant that the redbreast, because its food is more constant and plentiful,—seeds of grasses being preserved during the winter, and our farm-yards and stubble-fields furnishing an almost inexhaustible supply. Why, as a general rule, are aquatic, and especially sea birds, very numerous in individuals? Not because they are more prolific than others, generally the contrary; but because their food never fails, the sea-shores and river-banks daily swarming with a fresh supply of small mollusca and crustacea. Exactly the same laws will apply to mammals. Wild cats are prolific and have few enemies; why then are they never as abundant as rabbits? The only intelligible answer is, that their supply of food is more precarious. It appears evident, therefore, that so long as a country remains physically unchanged, the numbers of its animal population cannot materially increase. If one species does so, some others requiring the same kind of food must diminish in proportion. The numbers that die annually must be immense; and as the individual existence of each animal depends upon itself, those that die must be the weakest—the very young, the aged, and the diseased,—while those that prolong their existence can only be the most perfect in health and vigour—those who are best able to obtain food regularly, and avoid their numerous enemies. It is, as we commenced by remarking, "a struggle for existence," in which the weakest and least perfectly organized must always succumb.

Now it is clear that what takes place among the individuals of a species must also occur among the several allied species of a group—viz. that those which are best adapted to obtain a regular supply of food, and to defend themselves against the attacks of their enemies and the vicissitudes of the seasons, must necessarily obtain and preserve a superiority in population; while those species which from some defect of power or organization are the least capable of counteracting the vicissitudes of food, supply, etc., must diminish in numbers, and, in extreme cases, become altogether extinct. Between these extremes the species will present various degrees

[1] It is ironic and saddening that Wallace chose to illustrate this point about an abundant species, possessing strong survival characteristics, with the passenger pigeon, which is now extinct. *[The Editor.]*

of capacity for ensuring the means of preserving life; and it is thus we account for the abundance or rarity of species. Our ignorance will generally prevent us from accurately tracing the effects to their causes; but could we become perfectly acquainted with the organization and habits of the various species of animals, and could we measure the capacity of each for performing the different acts necessary to its safety and existence under all the varying circumstances by which it is surrounded, we might be able even to calculate the proportionate abundance of individuals which is the necessary result.

If now we have succeeded in establishing these two points—1st, *that the animal population of a country is generally stationary, being kept down by a periodical deficiency of food, and other checks*; and, 2nd, *that the comparative abundance or scarcity of the individuals of the several species is entirely due to their organization and resulting habits, which, rendering it more difficult to procure a regular supply of food and to provide for their personal safety in some cases than in others, can only be balanced by a difference in the population which have to exist in a given area*—we shall be in a condition to proceed to the consideration of *varieties*, to which the preceding remarks have a direct and very important application.

Most or perhaps all the variations from the typical form of a species must have some definite effect, however, slight, on the habits or capacities of the individuals. Even a change of colour might, by rendering them more or less distinguishable, affect their safety; a greater or less development of hair might modify their habits. More important changes, such as an increase in the power or dimensions of the limbs or any of the external organs, would more or less affect their mode of procuring food or the range of country which they inhabit. It is also evident that most changes would affect, either favourably or adversely, the powers of prolonging existence. An antelope with shorter or weaker legs must necessarily suffer more from the attacks of the feline carnivora; the passenger pigeon with less powerful wings would sooner or later be affected in its powers of procuring a regular supply of food; and in both cases the result must necessarily be a diminution of the population of the modified species. If, on the other hand, any species should produce a variety having slightly increased powers of preserving existence, that variety must inevitably in time acquire a superiority in numbers. These results must follow as surely as old age, intemperance, or scarcity of food produce an increased mortality. In both cases there may be many individual exceptions; but on the average the rule will invariably be found to hold good. All varieties will therefore fall into two classes—those which under the same conditions would never reach the population of the parent species, and those which would in time obtain and keep a numerical superiority. Now, let some alteration of physical conditions occur in the district—a long period of drought, a destruction of vegetation by locusts, the irruption of some new carnivorous animal seeking "pastures new"—any change in fact tending to render existence more difficult to the species in question, and tasking its utmost powers to avoid complete extermination; it is evident that, of all the individuals composing the species, those forming the least numerous and most feebly organized variety would suffer first, and, were the pressure severe, must soon become extinct. The same causes continuing in action, the parent species would next suffer, would gradually diminish in numbers, and with a recurrence of similar unfavourable conditions might also become extinct. The superior variety would then alone remain, and on a return to favourable circumstances would rapidly increase in numbers and occupy the place of the extinct species and variety.

The *variety* would now have replaced the *species*, of which it would be a more perfectly developed and more highly organized form. It would be in all respects better adapted to secure its safety, and to prolong its individual existence and that of the race. Such a variety *could not* return to the original form; for that form is an inferior one, and could never compete with it for existence. Granted, therefore, a "tendency" to reproduce the original type of the species, still the variety must ever remain preponderant in numbers, and under adverse physical conditions *again alone survive.* But this new, improved, and populous race might itself, in course of time, give rise to new varieties, exhibiting several diverging modifications of form, any of which, tending to increase the facilities for preserving existence, must by the same general law, in their turn become predominant. Here, then, we have *progres-*

sion and continued divergence deduced from the general laws which regulate the existence of animals in a state of nature, and from the undisputed fact that varieties do frequently occur. It is not, however, contended that this result would be invariable; a change of physical conditions in the district might at times materially modify it, rendering the race which had been the most capable of supporting existence under the former conditions now the least so, and even causing the extinction of the newer and, for a time, superior race, while the old or parent species and its first inferior varieties continued to flourish. Variations in unimportant parts might also occur, having no perceptible effect on the life-preserving powers; and the varieties so furnished might run a course parallel with the parent species, either giving rise to further variations or returning to the former type. All we argue for is, that certain varieties have a tendency to maintain their existence longer than the original species, and this tendency must make itself felt; for though the doctrine of chances or averages can never be trusted to on a limited scale, yet, if applied to high numbers, the results come nearer to what theory demands, and, as we approach to an infinity of examples, become strictly accurate. Now the scale on which nature works is so vast—the numbers of individuals and periods of time with which she deals approach so near to infinity, that any cause, however slight, and however liable to be veiled and counteracted by accidental circumstances, must in the end produce its full legitimate results.

Let us now turn to domesticated animals, and inquire how varieties produced among them are affected by the principles here enunciated. The essential difference in the condition of wild and domestic animals is this,—that among the former, their well-being and very existence depend upon the full exercise and healthy condition of all their senses and physical powers, whereas, among the latter, these are only partially exercised, and in some cases are absolutely unused. A wild animal has to search, and often to labour, for every mouthful of food—to exercise sight, hearing, and smell in seeking it, and in avoiding dangers, in procuring shelter from the inclemency of the seasons, and in providing for the subsistence and safety of its offspring. There is no muscle of its body that is not called into daily and hourly activity; there is no sense or faculty that is not strengthened by continual exercise. The domestic animal, on the other hand, has food provided for it, is sheltered, and often confined, to guard it against the vicissitudes of the seasons, is carefully secured from the attacks of its natural enemies, and seldom even rears its young without human assistance. Half of its senses and faculties are quite useless; and the other half are but occasionally called into feeble exercise, while even its muscular system is only irregularly called into action.

Now when a variety of such an animal occurs, having increased power or capacity in any organ or sense, such increase is totally useless, is never called into action, and may even exist without the animal ever becoming aware of it. In the wild animal, on the contrary, all its faculties and powers being brought into full action for the necessities of existence, any increase becomes immediately available, is strengthened by exercise, and must even slightly modify the food, the habits, and the whole economy of the race. It creates as it were a new animal, one of superior powers, and which will necessarily increase in numbers and outlive those inferior to it.

Again, in the domesticated animal all variations have an equal chance of continuance; and those which would decidedly render a wild animal unable to compete with its fellows and continue its existence are no disadvantage whatever in a state of domesticity. Our quickly fattening pigs, short-legged sheep, pouter pigeons, and poodle dogs could never have come into existence in a state of nature, because the very first step towards such inferior forms would have led to the rapid extinction of the race; still less could they now exist in competition with their wild allies. The great speed but slight endurance of the race horse, the unwielding strength of the ploughman's team, would both be useless in a state of nature. If turned wild on the pampas, such animals would probably soon become extinct, or under favorable circumstances might each lose those extreme qualities which would never be called into action, and in a few generations would revert to a common type, which must be that in which the various powers and faculties are so proportioned to each other as to be best adapted to procure food and secure safety,—that in which by the full exercise of every part of his organization the animal can alone continue to live. Domestic varieties, when turned wild, *must* return to something near the type of the original wild stock, or *become altogether extinct.*

We see, then, that no inferences as to varieties in a state of nature can be deduced from the observation of those occurring among domestic animals. The two are so much opposed to each other in every circumstance of their existence, that what applies to the one is almost sure not to apply to the other. Domestic animals are abnormal, irregular, artificial; they are subject to varieties which never occur and never can occur in a state of nature; their very existence depends altogether on human care: so far are many of them removed from that just proportion of faculties, that true balance of organization, by means of which alone an animal left to its own resources can preserve its existence and continue its race.

The hypothesis of Lamarck—that progressive changes in species have been produced by the attempts of animals to increase the development of their own organs, and thus modify their structure and habits—has been repeatedly and easily refuted by all writers on the subject of varieties and species, and its seems to have been considered that when this was done the whole question has been finally settled; but the view here developed renders such an hypothesis quite unnecessary, by showing that similar results must be produced by the action of principles constantly at work in nature. The powerful retractile talons of the falcon and the cat-tribes have not been produced or increased by the volition of those animals; but among the different varieties which occurred in the earlier and less highly organized forms of these groups, *those always survived longest which had the greatest facilities for seizing their prey.* Neither did the giraffe acquire its long neck by desiring to reach the foliage of the more lofty shrubs, and constantly stretching its neck for the purpose, but because any varieties which occurred among its antitypes with a longer neck than usual *at once secured a fresh range of pasture over the same ground as their shorter-necked companions, and on the first scarcity of food were thereby enabled to outlive them.* Even the peculiar colours of many animals, especially insects, so closely resembling the soil or the leaves or the trunks on which they habitually reside, are explained on the same principle; for though in the course of ages varieties of many tints may have occurred, *yet those races having colours best adapted to concealment from their enemies would inevitably survive the longest.* We have also here an acting cause to account for that balance so often observed in nature,—a deficiency in one set of organs always being compensated by an increased development of some others—powerful wings accompanying weak feet, or great velocity making up for the absence of defensive weapons; for it has been shown that all varieties in which an unbalanced deficiency occurred could not long continue their existence. The action of this principle is exactly like that of the centrifugal governor of the steam engine, which checks and corrects any irregularities almost before they become evident; and in like manner no unbalanced deficiency in the animal kingdom can ever reach any conspicuous magnitude, because it would make itself felt at the very first step, by rendering existence difficult and extinction almost sure to follow. An origin such as is here advocated will also agree with the peculiar character of the modifications of form and structure which obtain in organized beings—the many lines of divergence from a central type, the increasing efficiency and power of a particular organ through a succession of allied species, and the remarkable persistence of unimportant parts such as colour, texture of plumage and hair, form of horns or crests, through a series of species differing considerably in more essential characters. It also furnishes us with a reason for that "more specialized structure" which Professor Owen states to be a characteristic of recent compared with extinct forms, and which would evidently be the result of the progressive modification of any organ applied to a special purpose in the animal economy.

We believe we have now shown that there is a tendency in nature to the continued progression of certain classes of *varieties* further and further from the original type—a progression to which there appears no reason to assign any definite limits—and that the same principle which produces this result in a state of nature will also explain why domestic varieties have a tendency to revert to the original type. This progression, by minute steps, in various directions, but always checked and balanced by the necessary conditions, subject to which alone existence can be preserved, may, it is believed, be followed out so as to agree with all the phenomena presented by organized beings, their extinction and succession in past ages, and all the extraordinary modifications of form, instinct, and habits which they exhibit.

Ternate, February, 1858.

Journal of the Linnean Society, Zoology (1858) III, 45-62.

Questions and Topics for Discussion and Writing

1. What is the importance, for evolutionary theory, of the existence of "varieties" within a species (i.e. animals or plants belonging to a single species, but having certain different characteristics such as size, shape, colour, etc., of various parts of the body)?
2. What point is Wallace making when he discusses the reproductive rates of animals?
3. Explain the "struggle for existence."
4. How is it possible that Wallace, who knew nothing of Darwin's evolutionary ideas, came to virtually the identical theory of evolution?

The Scopes Trial

State of Tennessee

In 1925, the "Butler Act" was passed in the state of Tennessee, prohibiting the teaching of evolution in the public schools. To test the law before the courts, a high school science teacher named John Thomas Scopes agreed to be the defendant. He got himself arrested and indicted and stood trial in July of that year. The American Civil Liberties Union provided for his defence, led by the famous trial lawyer, Clarence Darrow. The prosecution lawyers included former presidential candidate William Jennings Bryan.

The trial riveted the attention of America on Tennessee. After Scopes' conviction Mississippi and Arkansas followed suit passing laws similar to the Butler Act. The Butler Act was finally repealed in 1967 and in 1968 the United States Supreme Court declared the Arkansas law unconstitutional.

The text which follows is excerpted from the official transcript of *State of Tennessee* vs. *John Thomas Scopes*. The trial went on for eleven days, so the transcript is quite long. These excerpts give some of the more memorable arguments on both sides, at the expense of some abrupt jumps to different parts of the transcript. The attorneys for the prosecution quoted here are William Jennings Bryan and General A. T. Stewart, the Attorney General of Tennessee. Counsel for the defence included Clarence Darrow, John R. Neal, Dudley Field Malone, and Arthur Garfield Hays.

GEN. STEWART—[*Reading*:] State of Tennessee, County of Rhea. Circuit Court. July Special Term, 1925.

The grand jurors for the state aforesaid, being duly summoned, elected, empanelled, sworn, and charged to inquire for the body of the county aforesaid, upon their oaths present:

That John Thomas Scopes, heretofore on the 24th day of April, 1925, in the county aforesaid, then and there, unlawfully did willfully teach in the public schools of Rhea County, Tennessee, which said public schools are supported in part and in whole by the public school fund of the state, a certain theory and theories that deny the story of the divine creation of man as taught in the Bible, and did teach instead thereof that man has descended from a lower order of animals, he, the said John Thomas Scopes, being at the time, and prior thereto, a teacher in the public schools of Rhea County, Tennessee, aforesaid, against the peace and dignity of the State.

THE COURT—What is your plea, gentlemen?

* * *

MR. NEAL—[*Reading*:] The defendant moves the court to quash the indictment in this case for the following reasons:

First—(A) Because the act which is the basis of the indictment, and which the defendant is charged with violating, is unconstitutional and void in that it violates...Section 3, Article I of the constitution of Tennessee:

> Section 3. Right of Worship Free—That all men have a natural and indefeasible right to worship Almighty God according to the dictates of his own conscience; that no man can of right, be compelled to attend, erect or support any place of worship, or to maintain any minister against his consent; that no human authority can, in any case whatever, control or interfere with the rights of conscience; and that no preference shall ever be given, by law, to any religious establishment or mode of worship.

* * *

GEN. STEWART—The next one, and the one which Dr. Neal referred to as one of the most important ones, Section 3, Article I, still of the constitution, the right of free worship...

If your Honor please, this law is as far removed from that interference with the provision in the constitution as it is from any other that is not even cited. This does not interfere with the religious worship—it does not even approach interference with religious worship...

MR. DARROW—...The part we claim is that last clause, "no preference shall ever be given, by law, to any religious establishment or mode of worship."

GEN. STEWART—Yes, that "no preference shall ever be given, by law, to any religious establishment or mode of worship." Then, how could that interfere, Mr. Darrow?

MR. DARROW—That is the part we claim is affected.

GEN. STEWART—In what wise?

MR. DARROW—Giving preference to the Bible.

* * *

GEN. STEWART—...There is as little in that as in any of the rest. If your Honor please, the St. James Version of the Bible is the recognized one in this section of the country. The laws of the land recognize the Bible; the laws of the land recognize the law of God and Christianity as a part of the common law.

MR. MALONE—Mr. Attorney General, may I ask a question?

GEN. STEWART—Certainly.

MR. MALONE—Does the law of the land or the law of the state of Tennessee recognize the Bible as part of a course in biology or science?

GEN. STEWART—I do not think the law of the land recognizes them as confusing one another in any particular.

* * *

GEN. STEWART—...The question involved here is, to my mind, the question of the exercise of the police power.

MR. NEAL—It does not mention the Bible?

GEN. STEWART—Yes, it mentions the Bible. The legislature, according to our laws, in my opinion, would have the right to preclude the teaching of geography. That is—

MR. NEAL—Does not it prefer the Bible to the Koran?

GEN. STEWART—It does not mention the Koran.

MR. MALONE—Does not it prefer the Bible to the Koran?

GEN. STEWART—We are not living in a heathen country.

MR. MALONE—Will you answer my question? Does not it prefer the Bible to the Koran?

GEN. STEWART—We are not living in a heathen country, so how could it prefer the Bible to the Koran?...

MR. MALONE—...I would say to base a theory set forth in any version of the Bible to be taught in the public school is an invasion of the rights of the citizen, whether exercised by the police power or by the legislature.

GEN. STEWART—Because it imposes a religious opinion?

MR. MALONE—Because it imposes a religious opinion, yes. What I mean is this: If there be in the state of Tennessee a single child or young man or young woman in your school who is a Jew, to impose upon any course of science a particular view of creation from the Bible is interfering, from our point of view, with his civil rights under our theory of the case. That is our contention.

* * *

MR. DARROW—...This case we have to argue is a case at law, and hard as it is for me to bring my mind to conceive it, almost impossible as it is to put my mind back into the sixteenth century, I am going to argue it as if it was serious, and as if it was a death struggle between two civilizations.

...We have been informed that the legislature has the right to prescribe the course of study in the public schools. Within reason, they no doubt have, no doubt. They could not prescribe it, I am inclined to think, under your constitution, if it omitted arithmetic and geography and writing, neither under the rest of the constitution, if it shall remain in force in the State, could they prescribe it if the course of study was only to teach religion, because several hundred years ago, when our people believed in freedom, and when no man felt so sure of their own sophistry that they were willing to send a man to jail who did not believe them, the people of Tennessee adopted a constitution, and they made it broad and plain, and said that the people of Tennessee would always enjoy religious freedom in its broadest terms; so, I assume, that no legislature could fix a course of study which violated that...

I remember, long ago, Mr. Bancroft wrote this sentence, which is true: "That it is all right to preserve freedom in constitutions, but when the spirit of freedom has fled from the hearts of the people, then its matter is easily sacrificed under law." And so it is, unless there is left enough of the spirit of freedom in the state of Tennessee, and in the United States, there is not a single line of any constitution that can withstand bigotry and ignorance when it seeks to destroy the rights of the individual; and bigotry and ignorance are ever active. Here, we find today as brazen and as bold an attempt to destroy learning as was ever made in the Middle Ages, and the only difference is we have not provided that they shall be burned at the stake, but there is time for that, your Honor; we have to approach these things gradually.

Now, let us see what we claim with reference to this law. If this proceeding, both in form and substance, can prevail in this court, then your Honor, no law—no matter haw foolish, wicked, ambiguous, or ancient, but can come back to Tennessee. All the guarantees go for nothing. All of the past has gone, will be forgotten, if this can succeed...

The statue should be comprehensible. It should not be written in Chinese anyway. It should be in passing English, as you say, so that common, human beings would understand what it meant, and so a man would know whether he is liable to go to jail when he is teaching, not so ambiguous as to be a snare or a trap to get someone who does not agree with you. If should be plain, simple and easy. Does this statute state what you shall teach and what you shall not? Oh, no! Not at all. Does it say you cannot teach the earth is round, because Genesis says it is flat? No. Does it say you cannot teach that the earth is millions of ages old, because the account in Genesis makes it less than six thousand years old? Oh, no. It doesn't state that. If it did you could understand it. It says you shan't teach any theory of the origin of man that is contrary to the divine theory contained in the Bible.

Now let us pass up the word "divine"! No legislature is strong enough in any state in the Union to characterize and pick any book as being divine. Let us take it as it is. What is the Bible?...The Bible is not one book. The Bible is made up of sixty-six books written over a period of about one thousand years, some of them very early and some of them comparatively late. It is a book primarily of religion and morals. It is not a book of science. Never was and was never meant to be. Under it there is nothing prescribed that would tell you how to build a railroad or a steamboat or to make anything that would advance civilization. It is not a textbook or a text on chemistry. It is not big enough to be. It is not a book on geology; they knew nothing about it. It is not a work on evolution; that is a mystery. It is not a work on astronomy. The man who looked out at the universe and studied the heavens had no thought but that the earth was the center of the universe. But we know better than that. We know that the sun is the center of the solar system. And that there are an infinity of other systems around about us. They thought the sun went around the earth and gave us night. We know better. We know the earth turns on its axis to produce days and nights. They thought the earth was [created] 4,004 years before the Christian Era. We know better. They told it the best they knew. And while suns may change all you may learn of chemistry, geometry and mathematics, there are no doubt certain primitive, elemental instincts in the organs of man that remain the same; he finds out what he can and yearns to know more and supplements his knowledge with hope and faith.

That is the province of religion and I haven't the slightest fault to find with it. Not the slightest in the world. One has one thought and one another, and instead of fighting each other as in the past, they should support and help each other. Let's see now. Can your Honor tell what is given as the origin of man as shown in the Bible? Is there any human being who can tell us? There are two conflicting accounts in the first two chapters. There are scattered all through it various acts and ideas, but to pass that up for the sake of argument, no teacher in any school in the state of Tennessee can know that he is violating a law, but must test every one of its doctrines by the Bible, must he not? You cannot say two times two equals four or make a man an educated man if evolution is forbidden. It does not specify what you cannot teach, but says you cannot teach anything that conflicts with the Bible. Then just imagine making it a criminal code that is so uncertain and impossible that every man must be sure that he has read everything in the Bible and not only read it but understands it, or he might violate the criminal code. Who is the chief mogul that can tell us what the Bible means? He or they should write a book and make it plain and distinct, so we would know. Let us look at it. There are in America at least five hundred different sects or churches, all of which quarrel with each other on the importance and nonimportance of certain things or the construction of certain passages. All along the line they do not agree among themselves and cannot agree among themselves. They never have and probably never will. There is a great division between the Catholics and the Protestants. There is such a disagreement that my client, who is a school-teacher, not only must know the subject he is teaching, but he must know everything about the Bible in reference to evolution. And he must be sure that he expresses it right or else some fellow will come along here, more ignorant perhaps than he, and say, "You made a bad guess and I think you have committed a crime." No criminal statute can rest that way...

It cannot stand a minute in this court on any theory than that it is a criminal act, simply because they say it contravenes the teaching of Moses, without telling us what those teachings are. Now, if this is the subject of a criminal act, then it cannot make a criminal out of a teacher in the public schools and leave a man free to teach it in a private school. It cannot make it criminal for a teacher in the public schools to teach evolution, and for the same man to stand among the hustings and teach it. It cannot make it a criminal act for this teacher to teach evolution and permit books upon evolution to be sold in every store in the state of Tennessee and to permit the newspapers from foreign cities to bring into your peaceful community the horrible utterances of evolution. Oh, no, nothing like that. If the state of Tennessee has any force in this day of Fundamentalism, in this day when religious bigotry and hatred is being kindled all over our land, see what can be done...

...Ignorance and fanaticism is ever busy and needs feeding. Always it is feeding and gloating for more. Today it is the public school teachers, tomorrow the private. The next day the preachers and the lecturers, the magazines, the books, the newspapers. After a while, your Honor, it is the setting of man against man and creed against creed, until with flying banners and beating drums we are marching backward to the glorious ages of the sixteenth century, when bigots lighted fagots to burn the men who dared to bring any intelligence and enlightenment and culture to the human mind...

MR. BRYAN—Little Howard Morgan—and, your Honor, that boy is going to make a great lawyer someday. I didn't realize it until I saw how a 14–year–old boy understood the subject so much better than a distinguished lawyer who attempted to quiz him. The little boy understood what he was talking about and, to my surprise, the attorneys didn't seem to catch the significance of the theory of evolution and the thought—and I'm sure he wouldn't have said it if he hadn't had thought it—he thought that little boy was talking about the individuals coming up from one cell. That wouldn't be evolution—that is growth, and one trouble about evolution is that it has been used in so many different ways that people are confused about it...

Tell me that the parents of this day have not any right to declare that children are not to be taught this doctrine? Shall not be taken down from the high plane upon which God put man? Shall be detached from the throne of God and be compelled to link their ancestors with the jungle; tell that to these children? Why, my friends, if they believe it, they go back to scoff at the religion of their parents! And the parents have a right to say that no teacher paid by their money shall rob their children of faith in God and send them back to their homes, skeptical, infidels, or agnostics, or atheists!

This doctrine that they want taught, this doctrine that they would force upon the schools, where they will not let the Bible be read—why, up in the state of New York they are now trying to keep the schools from adjourning for one hour in the afternoon, not that any teacher shall teach them the Bible, but that the children may go to the churches to which they belong and there have instruction in the Word. And they are refusing to let the school do that. These lawyers who are trying to force Darwinism and evolution on your children do not go back to protect the children of New York in their right to even have religion taught to them outside of the schoolroom, and they want to bring their experts in here...

Now, my friends, I want you to know that they not only have no proof, but they cannot find the beginning. I suppose this distinguished scholar who came here shamed them all by his number of degrees. He did not shame me, for I have more than he has, but I can understand how my friends felt when he unrolled degree after degree. Did he tell you where life began? Did he tell you that back of all these that there was a God? Not a word about it. Did he tell you how life began? Not a word, and not one of them can tell you how life began. The atheists say it came some way without a God; the agnostics say it came in some way, they know not whether with a God or not. And the Christian evolutionists say we came away back there somewhere, but they do not know how far back—they do not give you the beginning—not that gentleman that tried to qualify as an expert; he did not tell you whether it began with God or how. No, they take up life as a mystery that nobody can explain, and they want you to let them commence there and ask no questions. They want to come in with their little padded up evolution that commences with nothing and ends nowhere. They do not dare to tell you that it began with God and...ended with God. They come here with this bunch of stuff that they call evolution, that they tell you that everybody believes in, but do not know that everybody knows as a fact, and nobody can tell how it came, and they do not explain the great riddle of the universe—they do not deal with the problems of life—they do not teach the great science of how to live—and yet they would undermine the faith of these little children in that God who stands back of everything and whose promise we have that we shall live with Him forever bye and bye. They shut God out of the world. They do not talk about God. Darwin says the beginning of all things is a mystery unsolvable by us. He does not pretend to say how these things started...

And your Honor asked me whether it has anything to do with the principle of the virgin birth. Yes, because this principle of evolution disputes the miracle; there is no place for the miracle in this train of evolution, and the Old Testament and the New are filled with miracles, and if this doctrine is true, this logic eliminates every mystery in the Old Testament and the New, and eliminates everything supernatural; and that means they eliminate the virgin birth—that means that they eliminate the resurrection of the body—that means that they eliminate the doctrine of atonement. And they believe man has been rising all the time, that man never fell; that when the Savior came there was not any reason for His coming; there was no reason why He should not go as soon as He could, that He was born of Joseph or some other correspondent, and that He lies in his grave. And when the Christians of this state have tied their hands and said, "We will not take advantage of our power to teach religion to our children, by teachers paid by us," these people come in from the outside of the state and force upon the people of this state and upon the children of the taxpayers of this state a doctrine that refutes not only their belief in God, but their belief in a Savior and belief in heaven, and takes from them every moral standard that the Bible gives us...

Your Honor, we first pointed out that we do not need any experts in science....And, when it comes to Bible experts, every member of the jury is as good an expert on the Bible as any man that they could bring, or that we could bring. The one beauty about the Word of God is, it does not take an expert to understand it. They have translated that Bible into five hundred languages; they have carried it into nations where but few can read a word, or write, to people who never saw a book, who never read, and yet can understand that Bible, and they can accept the salvation that the Bible offers, and they can know more about that book by accepting Jesus and feeling in their hearts the sense of their sins forgiven than all of the skeptical outside Bible experts that could come in here to talk to the people of Tennessee about the construction that they place upon the Bible, that is foreign to the construction that the people here place upon it.

MR. HAYS—The Defense desires to call Mr. Bryan as a witness, and, of course, since the only question here is whether Mr. Scopes taught what these children said he taught, we recognize what Mr. Bryan says as a witness would not be very valuable. We think there are other questions involved, and we should want to take Mr. Bryan's testimony for the purposes of our record, even if your Honor thinks it is not admissible in general, so we wish to call him now.

THE COURT—Mr. Bryan, you are not objecting to going on the stand?

MR. BRYAN—Not at all.

THE COURT—Do you want Mr. Bryan sworn?

MR. DARROW—no.

MR. BRYAN—I can make affirmation; I can say, "So help me God, I will tell the truth."

MR. DARROW—No, I take it you will tell the truth, Mr. Bryan.

Q [MR. DARROW]—You have given considerable

study to the Bible, haven't you, Mr. Bryan?

A [MR. BRYAN]—Yes, sir, I have tried to.

* * *

Q—Do you claim that everything in the Bible should be literally interpreted?

A—I believe everything in the Bible should be accepted as it is given there; some of the Bible is given illustratively. For instance: "Ye are the salt of the earth." I would not insist that man was actually salt, or that he had flesh of salt, but it is used in the sense of salt as saving God's people.

Q—But when you read that Jonah swallowed the whale—or that the whale swallowed Jonah—excuse me please—how do you literally interpret that?

A—When I read that a big fish swallowed Jonah—it does not say whale.

Q—Doesn't it? Are you sure?

A—That is my recollection of it. A big fish, and I believe it; and I believe in a God who can make a whale and can make a man and make both do what He pleases.

Q—Mr. Bryan, doesn't the New Testament say a whale?

A—I am not sure. My impression is that it says fish; but it does not make so much difference; I merely called your attention to where it says fish—it does not say whale.

Q—But in the New Testament it says whale, doesn't it?

A—That may be true; I cannot remember in my own mind what I read about it.

Q—Now, you say, the big fish swallowed Jonah, and he there remained how long? three days? and then he spewed him upon the land. You believe that the big fish was made to swallow Jonah?

A—I am not prepared to say that; the Bible merely says it was done.

Q—You don't know whether it was the ordinary run of fish, or made for that purpose?

A—You may guess; you evolutionists guess.

Q—But when we do guess, we have a sense to guess right.

A—But do not do it often.

Q—You are not prepared to say whether that fish was made especially to swallow a man or not?

A—The Bible doesn't say, so I am not prepared to say.

Q—You don't know whether that was fixed up specially for the purpose?

A—No, the Bible doesn't say.

Q—But you do believe He made them—that He made such a fish and that it was big enough to swallow Jonah?

A—Yes, sir. Let me add: one miracle is just as easy to believe as another.

Q—It is for me.

A—It is for me.

Q—Just as hard?

A—It is hard to believe for you, but easy for me. A miracle is a thing performed beyond what man can perform. What you get beyond what man can do, you get with the realm of miracles; and it is just as easy to believe the miracle of Jonah as any other miracle in the Bible.

Q—Perfectly easy to believe that Jonah swallowed the whale?

A—If the Bible said so; the Bible doesn't make as extreme statements as evolutionists do.

MR. DARROW—That may be a question, Mr. Bryan, about some of those you have known.

A—The only thing is, you have a definition of fact that includes imagination.

Q—And you have a definition that excludes everything but imagination.

GEN. STEWART—I object to that as argumentative.

THE WITNESS—You—

MR. DARROW—The witness must not argue with me, either.

Q—Do you consider the story of Jonah and the whale a miracle?

A—I think it is.

Q—Do you believe Joshua made the sun stand still?

A—I believe what the Bible says. I suppose you mean that the earth stood still?

Q—I don't know. I am talking about the Bible now.

A—I accept the Bible absolutely.

Q—The Bible says Joshua commanded the sun to stand still for the purpose of lengthening the day, doesn't it? and you believe it?

A—I do.

Q—Do you believe at that time the entire sun went around the earth?

A—No, I believe that the earth goes around the sun.

State of Tennessee

Q—Do you believe that men who wrote it thought that the day could be lengthened or that the sun could be stopped?

A—I don't know what they thought.

Q—You don't know?

A—I think they wrote the fact without expressing their own thoughts.

* * *

MR. DARROW—Have you an opinion as to whether—whoever wrote the book, I believe it is, Joshua, the Book of Joshua, thought the sun went around the earth or not?

A—I believe that he was inspired.

MR. DARROW—Can you answer my question?

A—When you let me finish the statement.

Q—It is a simple question, but finish it.

THE WITNESS—You cannot measure the length of my answer by the length of your question.

[*Laughter in the courtyard.*]

MR. DARROW—No, except that the answer be longer.

[*Laughter in the courtyard.*]

A—I believe that the Bible is inspired, an inspired author, whether one who wrote as he was directed to write understood the things he was writing about, I don't know.

Q—Whoever inspired it? Do you think whoever inspired it believed that the sun went around the earth?

A—I believe it was inspired by the Almighty, and He may have used language that could be understood at that time.

Q—Was—

THE WITNESS—Instead of using language that could not be understood until Mr. Darrow was born.

[*Laughter and applause in the courtyard.*]

* * *

Q—You believe the story of the flood to be a literal interpretation?

A—Yes, sir.

Q—When was that flood?

A—I would not attempt to fix the date. The date is fixed, as suggested this morning.

Q—About 4004 B.C.?

A—That has been the estimate of a man that is accepted today. I would not say it is accurate.

* * *

MR. DARROW—How long ago was the flood, Mr. Bryan?

* * *

THE WITNESS—Oh, I would put the estimate where it is, because I have no reason to vary it. But I would have to look at it to give you the exact date.

Q—I would, too. Do you remember what book the account is in?

A—Genesis.

MR. HAYS—Is that the one in evidence?

MR. NEAL—That will have it; that is the King James Version.

MR. DARROW—The one in evidence has it.

THE WITNESS—It is given here, as 2348 years B.C.

Q—Well, 2348 years B.C. You believe that all the living things that were not contained in the ark were destroyed.

A—I think the fish may have lived.

Q—Outside of the fish?

A—I cannot say.

Q—You cannot say?

A—No, I accept that just as it is; I have no proof to the contrary.

Q—I am asking you whether you believe?

A—I do.

Q—That all living things outside of the fish were destroyed?

A—What I say about the fish is merely a matter of humor.

Q—I understand.

* * *

Q—Don't you know that the ancient civilizations of China are 6,000 or 7,000 years old, at the very least?

A—No; but they would not run back beyond the creation, according to the Bible, 6,000 years.

Q—You don't know how old they are, is that right?

A—I don't know how old they are, but probably you do. [*Laughter in the courtyard.*] I think you would give preference to anybody who opposed the Bible, and I give the preference to the Bible.

Q—I see. Well, you are welcome to your opinion. Have you any idea how old the Egyptian civilization is?

A—No.

Q—Do you know of any record in the world, outside of the story of the Bible, which conforms to any statement that it is 4,200 years ago or thereabouts that all life was wiped off the face of the earth?

A—I think they have found records.

Q—Do you know of any?

A—Records reciting the flood, but I am not an authority on the subject.

Q—Now, Mr. Bryan, will you say if you know of any record, or have ever heard of any records, that describe that a flood existed 4,200 years ago, or about that time, which wiped all life off the earth?

A—The recollection of what I have read on that subject is not distinct enough to say whether the records attempted to fix a time, but I have seen in the discoveries of archaeologists where they have found records that described the flood.

Q—Mr. Bryan, don't you know that there are many old religions that describe the flood?

A—No, I don't know.

Q—You know there are others besides the Jewish?

A—I don't know whether these are the record of any other religion or refer to this flood.

Q—Don't you ever examine religion so far to know that?

A—Outside of the Bible?

Q—Yes.

A—No; I have not examined to know that, generally.

Q—You have never examined any other religions?

A—Yes, sir.

Q—Have you ever read anything about the origins of religions?

A—Not a great deal.

Q—You have never examined any other religion?

A—Yes, sir.

Q—And you don't know whether any other religion ever gave a similar account of the destruction of the earth by the flood?

A—The Christian religion has satisfied me, and I have never felt it necessary to look up some competing religions.

State of Tennessee vs. *John Thoman Scopes,* Nos. 5231 and 5232, in the Circuit Court of Rhea County, Tennessee.

Questions and Topics for Discussion and Writing

1. Why does Clarence Darrow compare the Scopes trial to the Inquisition of the sixteenth century?
2. What is Darrow trying to achieve by putting William Jennings Bryan—the *counsel* for the prosecution—on the witness stand and questioning him?
3. What is actually at issue in this trial (judging from the small excerpt at hand)? Is it the fact that Scopes taught evolution in the schools? Is it the truth or falsity of evolution? Is it the issue of whether evolution is in conflict with the Bible? Is it some other issue?

Darwinism Defined: The Difference Between Fact and Theory

Stephen Jay Gould

In this essay, Stephen Jay Gould, professor of biology, geology, and the history of science at Harvard University and a very popular writer on evolution, takes to task those commentators who confuse the fact of evolution with theories that explain it.

Charles Darwin, who was, perhaps, the most incisive thinker among the great minds of history, clearly divided his life's work into two claims of different character: establishing the fact of evolution, and proposing a theory (natural selection) for the mechanism of evolutionary change. He also expressed, and with equal clarity, his judgment about their different status: confidence in the facts of transmutation and genealogical connection among all organisms, and appropriate caution about his unproved theory of natural selection. He stated in the *Descent of Man*: "I had two distinct objects in view; firstly, to show that species had not been separately created, and secondly, that natural selection had been the chief agent of change...If I have erred in...having exaggerated its [natural selection's] power...I have at least, as I hope, done good service in aiding to overthrow the dogma of separate creations."

Darwin wrote those words more than a century ago. Evolutionary biologists have honored his fundamental distinction between fact and theory ever since. Facts are the world's data; theories are explanations proposed to interpret and coordinate facts. The fact of evolution is as well established as anything in science (as secure as the revolution of the earth about the sun), though absolute certainty has no place in our lexicon. Theories, or statements about the causes of documented evolutionary change, are now in a period of intense debate—a good mark of science in its healthiest state. Facts don't disappear while scientists debate theories. As I wrote in an early issue of this magazine (*Discover*, May 1981), "Einstein's theory of gravitation replaced Newton's, but apples did not suspend themselves in mid-air pending the outcome."

Since facts and theories are so different, it isn't surprising that these two components of science have had separate histories ever since Darwin. Between 1859 (the year of publication for the *Origin of Species*) and 1882 (the year of Darwin's death), nearly all thinking people came to accept the fact of evolution. Darwin lies beside Newton in Westminster Abbey for this great contribution. His theory of natural selection has experienced a much different, and checkered, history. It attracted some notable followers during his lifetime (Wallace in England, Weismann in Germany), but never enjoyed majority support. It became an orthodoxy among English-speaking evolutionists (but never, to this day, in France or Germany) during the 1930s, and received little cogent criticism until the 1970s. The past fifteen years have witnessed a revival of intense and, this time, highly fruitful debate as scientists discover and consider the implications of phenomena that expand the potential causes of evolution well beyond the unitary focus of strict Darwinism (the struggle for reproductive success among organisms within populations). Darwinian selection will not be overthrown; it will remain a central focus of more inclusive evolutionary theories. But new findings and interpretations at all levels, from molecular change in genes to patterns of overall diversity in geological time, have greatly expanded the scope of important causes—from random, selectively neutral change at the genetic level, to punctuated equilibria and catastrophic mass extinction in geological time.

In this period of vigorous pluralism and intense debate among evolutionary biologists, I am greatly saddened to note that some distinguished commentators among non-scientists, in particular Irving Kristol in a *New York Times* Op Ed piece[1] of Sept. 30, 1986 ("Room for Darwin and the Bible"), so egregiously misunderstand the character of our discipline and continue to confuse this central distinction between secure fact and healthy debate about theory.

I don't speak of the militant fundamentalists who label themselves with the oxymoron "scientific creationists," and try to sneak their Genesis literalism into high school classrooms under the guise of scientific dissent. I'm used to their rhetoric, their dishonest mis- and half-quotations, their constant repetition of "useful" arguments that even they must recognize as nonsense (disproved human footprints on dinosaur trackways in Texas, visible misinterpretation of thermodynamics to argue that life's complexity couldn't increase without a divine boost). Our struggle with these ideologues is political, not intellectual. I speak instead of our allies among people committed to reason and honorable argument.

Kristol, who is no fundamentalist, accuses evolutionary biologists of bringing their troubles with creationists upon themselves by too zealous an insistence upon the truths of Darwin's world. He writes: "...the debate has become a dogmatic crusade on both sides, and our educators, school administrators, and textbook publishers find themselves trapped in the middle." He places the primary blame upon a supposedly anti-religious stance in biological textbooks: "There is no doubt that most of our textbooks are still written as participants in the 'warfare' between science and religion that is our heritage from the 19th century. And there is also little doubt that it is this pseudoscientific dogmatism that has provoked the current religious reaction."

Kristol needs a history lesson if he thinks that current creationism is a product of scientific intransigence. Creationism, as a political movement against evolution, has been a continually powerful force since the days of the Scopes trial. Rather than using evolution to crusade against religion in their texts, scientists have been lucky to get anything at all about evolution into books for high school students ever since Scopes's trial in 1925. My own high school biology text, used in the liberal constituency of New York City in 1956, didn't even mention the word evolution. The laws that were used against Scopes and cowed textbook publishers into submission weren't overturned by the Supreme Court until 1968 (*Epperson v. Arkansas*).

But what about Kristol's major charge—anti-religious prejudice and one-dimensional dogmatism about evolution in modern textbooks? Now we come to the heart of what makes me so sad about Kristol's charges and others in a similar vein. I don't deny that some texts have simplified, even distorted, in failing to cover the spectrum of modern debates; this, I fear, is a limitation of the genre itself (and the reason why I, though more of a writer than most scientists, have never chosen to compose a text). But what evidence can Kristol or anyone else provide to demonstrate that evolutionists have been worse than scientists from other fields in glossing over legitimate debate within their textbooks?

Consider the evidence. Two textbooks of evolution now dominate the field. One has as its senior author Theodosius Dobzhansky, the greatest evolutionist of our century, and a lifelong Russian Orthodox; nothing anti-religious could slip past his watchful eye. The second, by Douglas Futuyma, is a fine book by a kind and generous man who could never be dogmatic about anything except intolerance. (His book gives a fair hearing to my own heterodoxies, while dissenting from them.)

When we come to popular writing about evolution, I suppose that my own essays are as well read as any. I don't think that Kristol could include me among Darwinian dogmatists, for most of my essays focus upon my disagreements with the strict version of natural selection. I also doubt that Kristol

[1] An invited editorial article published on the page opposite to the newspaper's own editorials—hence an "Op Ed" piece. *[The Editor.]*

would judge me anti-religious, since I have campaigned long and hard against the same silly dichotomy of science versus religion that he so rightly ridicules. I have written laudatory essays about several scientists (Burnet, Duvier, Buckland, and Gosse, among others) branded as theological dogmatists during the nineteenth-century reaction; and, while I'm not a conventional believer, I don't consider myself irreligious.

Kristol's major error lies in his persistent confusion of fact with theory. He accuses us—without giving a single concrete example, by the way—of dogmatism about *theory* and sustains his charge by citing our confidence in the *fact* of transmutation. "It is reasonable to suppose that if evolution were taught more cautiously, as a conglomerate idea consisting of conflicting hypotheses rather than as an unchallengeable certainty, it would be far less controversial."

Well, Mr. Kristol, evolution (as theory) is indeed "a conglomerate idea consisting of conflicting hypotheses," and I and my colleagues teach it as such. But evolution is also a fact of nature, and so do we teach it as well, just as our geological colleagues describe the structure of silicate minerals, and astronomers the elliptical orbits of planets.

Rather than castigate Mr. Kristol any further, I want to discuss the larger issue that underlies both this incident and the popular perception of evolution in general. If you will accept my premise that evolution is as well established as any scientific fact (I shall give the reasons in a moment), then why are we uniquely called upon to justify our chosen profession; and why are we alone subjected to such unwarranted infamy? To this central question of this essay, I suggest the following answer. We haven't received our due for two reasons: (1) a general misunderstanding of the different methods used by all historical sciences (including evolution), for our modes of inference don't match stereotypes of "*the* scientific method"; and (2) a continuing but unjustified fear about the implication both of evolution itself and of Darwin's theory for its mechanism. With these two issues resolved, we can understand both the richness of science (in its pluralistic methods of inquiry) and the absence of any conflict, through lack of common content, between proper science and true religion.

Our confidence in the fact of evolution rests upon copious data that fall, roughly, into three great classes. First, we have the direct evidence of small-scale changes in controlled laboratory experiments of the past hundred years (on bacteria, on almost every measurable property of the fruit fly *Drosophila*), or observed in nature (color changes in moth wings, development of metal tolerance in plants growing near industrial waste heaps), or produced during a few thousand years of human breeding and agriculture. Creationists can scarcely ignore this evidence, so they respond by arguing that God permits limited modification within created types, but that you can never change a cat into a dog (who ever said that you could, or that nature did?).

Second, we have direct evidence for large-scale changes, based upon sequences in the fossil record. The nature of this evidence is often misunderstood by non-professionals who view evolution as a simple ladder of progress, and therefore expect a linear array of "missing links." But evolution is a copiously branching bush, not a ladder. since our fossil record is so imperfect, we can't hope to find evidence for every tiny twiglet. (Sometimes, in rapidly evolving lineages of abundant organisms restricted to a small area and entombed in sediments with an excellent fossil record, we do discover an entire little bush—but such examples are as rare as they are precious.) In the usual case, we may recover the remains of side branch number 5 from the bush's early history, then bough number 40 a bit later, then the full series of branches 156–161 in a well preserved sequence of younger rocks, and finally surviving twigs 250 and 287.

In other words, we usually find sequences of structural intermediates, not linear arrays of ancestors and descendants. Such sequences provide superb examples of temporally ordered evolutionary trends. Consider the evidence for human evolution in Africa. What more could you ask from a record of rare creatures living in terrestrial environments that provide poor opportunity for fossilization? We have a temporal sequence displaying clear trends in a suite of features, including threefold increase of brain size and corresponding decrease of jaws and teeth. (We are missing direct evidence for an earlier transition to upright posture, but wide-ranging and unstudied sediments of the right age have been found in East Africa, and we have an excellent chance to fill in this part of our story.) What alternative can we suggest to evolution? Would God—for some inscrutable reason, or merely to test our faith—create five species, one after the other (*Australopithecus afarensis, A. Africanus, Homo habilis, H. erectus, and H. sapiens*), to mimic a continuous trend of evolutionary change?

Or, consider another example with evidence of structurally intermediate stages—the transition from reptiles to mammals. The lower jaw of mammals contains but a single bone, the dentary. Reptiles build their lower jaws of several bones. In perhaps the most fascinating of those quirky changes in function that mark pathways of evolution, the two bones articulating the upper and lower jaws of reptiles migrate to the middle ear and become the malleus and incus (hammer and anvil) of mammals.

Creationists, ignorant of hard evidence in the fossil record, scoff at this tale. How could jaw bones become ear bones, they ask. What happened in between? An animal can't work with a jaw half disarticulated during the stressful time of transition.

The fossil record provides a direct answer. In an excellent series of temporally ordered structural intermediates, the reptilian dentary gets larger and larger, pushing back as the other bones of a reptile's lower jaw decrease in size. We've even found a transitional form with an elegant solution to the problem of remaking jaw bones into ear bones. This creature has a double articulation—one between the two bones that become the mammalian hammer and anvil (the old reptilian joint), and a second between the squamosal and dentary bones (the modern mammalian condition). With this built-in redundancy, the emerging mammals could abandon one connection by moving two bones into the ear, while retaining the second linkage, which becomes the sole articulation of modern mammals.

Third, and most persuasive in its ubiquity, we have the signs of history preserved within every organism, every ecosystem, and every pattern of biogeographic distribution, by those pervasive quirks, oddities, and imperfections that record pathways of historical descent. These evidences are indirect, since we are viewing modern results, not the processes that caused them, but what else can we make of the pervasive pattern? Why does our body, from the bones of our back to the musculature of our belly, display the vestiges of an arrangement better suited for quadrupedal life if we aren't the descendants of four-footed creatures? Why do the plants and animals of the Galapagos so closely resemble, but differ slightly from the creatures of Equador, the nearest bit of land 600 miles to the east, especially when cool oceanic currents and volcanic substrate make the Galapagos such a different environment from Ecuador (thus removing the potential argument that God makes the best creatures for each place, and small differences only reflect a minimal disparity of environments)? The similarities can only mean that Ecudorian creatures colonized the Galapagos and then diverged by a natural process of evolution.

This method of searching for oddities as vestiges of the past isn't peculiar to evolution, but a common procedure of all historical science. How, for example, do we know that words have histories, and haven't been decreed by some all-knowing committee in Mr. Orwell's bureau of Newspeak? Doesn't the bucolic etymology of so many words testify to a different life style among our ancestors? In this article, I try to "broadcast" some ideas (a mode of sowing seed) in order to counter the most "egregious" of creationist sophistries (the animal *ex grege*, or outside the flock), for which, given the *quid pro quo* of business, this fine magazine pays me an "emolument" (the fee that millers once received to grind corn).

I don't want to sound like a shrill dogmatist shouting"rally round the flag boys," but biologists have reached a consensus, based on these kinds of data, about the fact of evolution. When honest critics like Irving Kristol misinterpret this agreement, they're either confusing our fruitful consonance about mechanisms of change, or they've misinterpreted part of our admittedly arcane technical literature.

Once such misinterpretation has gained sufficient notoriety in the last year that we crave resolution both for its own sake and as an illustration of the frustrating confusion that can arise when scientists aren't clear and when commentators, as a result of hidden agendas, don't listen. Tom Bethell argued in *Harper's* (February 1985) that a group of young taxonomists called pattern cladists have begun to doubt the existence of evolution itself.

This would be truly astounding news, since cladistics is a powerful method dedicated to reforming classification by using only the branching order of lineages on evolutionary trees ("propinquity of descent" in Darwin's lovely phrase), rather than vague notions of overall similarity in form or function. (For example, in the cladistic system, a lungfish is more closely related to a horse than to a salmon because the common ancestor of lungfish and horse is more recent in time than the link point of the lungfish-horse lineage with the branch leading to modern bony fishes (including salmon).

Cladists use only the order of branching to construct their schemes of relationships; it bothers them not a whit that lungfish and salmon look and work so much alike. Cladism, in other words, is the purest of all genealogical systems for classification, since it works only with closeness of common ancestry in time. How preciously ironic then, that this most rigidly evolutionary of all taxonomic systems should become the subject of such extraordinary misunderstanding—as devised by Bethell, and perpetuated by Kristol when he writes: "...many younger biologists (the so-called 'cladists') are persuaded that the differences among species—including those that seem to be closely related—are such as to make the very concept of evolution questionable."

This error arose for the following reason. A small splinter group of cladists (not all of them, as Kristol claims)—"transformed" or "pattern" cladists by their own designation—have adopted what is to me an ill-conceived definition of scientific procedure. They've decided, by misreading Karl Popper's philosophy, that patterns of branching can be established unambiguously as a fact of nature, but that processes causing events of branching, since they can't be observed directly, can't be known with certainty. Therefore, they say, we must talk only of pattern and rigidly exclude all discussion of process (hence "pattern cladistics").

This is where Bethell got everything arse-backwards and began the whole confusion. A philosophical choice to abjure all talk about process isn't the same thing as declaring that no reason for patterns of branching exists. Pattern cladists don't doubt that evolution is the cause behind branching; rather, they've decided that our science shouldn't be discussing causes at all.

Now I happen to think that this philosophy is misguided; in unguarded moments I would even deem it absurd. Science, after all, is fundamentally about process; learning why and how things happen is the soul of our discipline. You can't abandon the search for cause in favor of a dry documentation of pattern. You must take risks of uncertainty in order to probe the deeper questions, rather than stopping with sterile security. You see, now I've blown our cover. We scientists do have our passionate debates—and I've just poured forth an example. But as I wrote earlier, this is a debate about the proper approach to causes, not an argument about whether causes exist, or even whether the cause of branching is evolution or something else. No cladist denies that branching patterns arise by evolution.

This incident also raises the troubling issue of how myths become beliefs through adulterated repetition without proper documentation. Bethell began by misunderstanding pattern cladistics, but at least he reports the movement as a small splinter, and tries to reproduce their arguments. then Kristol picks up the ball and recasts it as a single sentence of supposed fact—and all cladists have now become doubters of evolution by proclamation. Thus a movement, by fiat, is turned into its opposite—as the purest of all methods for establishing genealogical connections becomes a weapon for denying the mechanism that all biologists accept as the cause of branching on life's tree: evolution itself. Our genealogy hasn't been threatened, but my geniality has almost succumbed.

When I ask myself why the evidence for evolution, so clear to all historical scientists, fails to impress intelligent nonscientists, I must believe that more than simple misinformation lies at the root of our difficulty with a man like Irving Kristol. I believe that the main problem centers upon a restrictive stereotype of scientific method accepted by most non-practitioners as the essential definition of all scientific work.

We learn in high school about *the* scientific method—a cut-and-dried procedure of simplication to essential components, experiment in the controlled situation of a laboratory, prediction and replication. But the sciences of history—not just evolution but a suite of fundamental disciplines ranging from geology, to cosmology, to linguistics—can't operate by this stereotype. We are charged with explaining events of extraordinary complexity that occur but once in all their details.

We try to understand the past, but don't pretend to predict the future. We can't see past processes directly, but learn to infer their operation from preserved results.

Science is a pluralistic enterprise with a rich panoply of methods appropriate for different kinds of problems. Past events of long duration don't lie outside the realm of science because we cannot make them happen in a month within our laboratory. Direct vision isn't the only, or even the usual, method of inference in science. We don't see electrons, or quarks, or chemical bonds, any more than we see small dinosaurs evolve into birds, or India crash into Asia to raise the Himalayas.

William Whewell, the great English philosopher of science during the early nineteenth century, argued that historical science can reach conclusions as well confirmed as any derived from experiment and replication in laboratories, by a method he called "consilience" (literally "jumping together") of inductions. Since we can't see the past directly or manipulate its events, we must use the different tactic of meeting history's richness head on. We must gather its wondrously varied results and search for a coordinating cause that can make sense of disparate data otherwise isolated and uncoordinated. We must see if a set of results so diverse that no one had ever considered their potential coordination might jump together as the varied products of a single process. Thus plate tectonics can explain magnetic stripes on the sea floor, the rise and later erosion of the Appalachians, the earthquakes of Lisbon and San Francisco, the eruption of Mount St. Helens, the presence of large flightless ground birds only on continents once united as Gondwanaland, and the discovery of fossil coal in Antarctica.

Darwin, who understood the different rigor of historical science so well, complained bitterly about those critics who denied scientific status to evolution because they couldn't see it directly or reproduce its historical results in a laboratory. He wrote to Hooker in 1861: "Change of species cannot be directly proved...The doctrine must sink or swim according as it groups and explains phenomena. It is really curious how few judge it in this way, which is clearly the right way." And later, in 1868: "This hypothesis may be tested...by trying whether it explains several large and independent classes of facts; such as the geological succession of organic beings, their distribution in past and present times, and their mutual affinities and homologies."

If a misunderstanding of the different methods of historical inquiry has impeded the recognition of evolution as a product of science at its best, then a residual fear for our own estate has continued to foster resentment of the fact that our physical bodies have ancient roots in ape-like primates, waddling reptiles, jawless fishes, worm-like invertebrates, and other creatures deemed even lower or more ignoble. Our ancient hopes for human transcendence have yet to make their peace with Darwin's world.

But what challenge can the facts of nature pose to our own decisions about the moral value of our lives? We are what we are, but we interpret the meaning of our heritage as we choose. Science can no more answer the questions of how we ought to live than religion can decree the age of the earth. Honorable and discerning scientists (most of us, I trust) have always understood that the limits to what science can answer also describe the power of its methods in their proper domain. Darwin himself exclaimed that science couldn't touch the problem of evil and similar moral conundrums: "A dog might as well speculate on the mind of Newton. Let each man hope and believe what he can."

There is no warfare between science and religion, never was except as a historical vestige of shifting taxonomic boundaries among disciplines. Theologians haven't been troubled by the fact of evolution, unless they try to extend their own domain beyond its proper border (hubris and territorial expansionism aren't the sins of scientists alone, despite Mr. Kristol's fears). The Reverend Henry Ward Beecher, our greatest orator during Darwin's century, evoked the most quintessential of American metaphors in dismissing the entire subject of conflict between science and religion with a single epithet: "Design by wholesale is grander than design by retail"—or, general laws rather than creation of each item by fiat will satisfy our notion of divinity.

Similarly, most scientists show no hostility to religion. Why should we, since our subject doesn't intersect the concerns of theology? I strongly dispute Kristol's claim that "the current teaching of evolution in our public schools does indeed have an ideological bias against religious belief." Unless at least half my colleagues are inconsistent dunces, there can be—on the most raw and direct empirical grounds—no conflict between science and religion. I know hundreds of scientists who share a conviction about the fact of evolution, and teach it in much the same way. Among these people I note an entire spectrum of religious attitudes—from devout daily prayer and worship to resolute atheism. Either there's no correlation between religious belief and confidence in evolution—or else half these people are fools.

The common goal of science and religion is our shared struggle for wisdom in all its various guises. I know no better illustration of this great unity than a final story about Charles Darwin. This scourge of fundamentalism had a conventional church burial—in Westminster Abbey no less. J. Frederick Bridge, Abbey organist and Oxford don, composed a funeral anthem especially for the occasion. It may not rank high in the history of music, but it is, as my chorus director opined, a "sweet piece." (I've made what may be the only extant recording of this work, marred only by the voice of yours truly within the bass section.) Bridge selected for this text the finest biblical description of the common aim that will forever motivate both the directors of his building and the inhabitants of the temple of science—wisdom. "Her ways are ways of pleasantness and all her paths are peace" (Proverbs 3: 17).

I am only sorry that Dr. Bridge didn't set the very next metaphor about wisdom (Proverbs 3: 18), for it describes, with the proper topology of evolution itself, the greatest dream of those who followed the God of Abraham, Isaac, and Jacob: "She is a tree of life to them that lay hold upon her."

From *Discover* (January, 1987), 64–70.

Questions and Topics for Discussion and Writing

1. Explain the distinction that Gould draws between evolution as a historical fact and the theory of evolution.
2. Why is it so incredible that cladists would be reported to have doubts about evolution?
3. Does Gould believe that science and religion are mutually opposed to each other? What does Gould think of scientific creationism?

The Discovery of Units of Heredity

Ruth Moore

The theory of evolution by natural selection as expounded by Darwin or Wallace had one very serious flaw: there was no adequate explanation of the mechanism of inheritance. Not only could Darwin not explain how species could slowly change characteristics, he could not really explain how species in general stayed the same from generation to generation. It was a problem that troubled all the theoretically minded evolutionists in Darwin's day, and drove some of them back toward accepting Lamarck's view that animals acquired characteristics during their own lifetimes and then passed on those new characteristics to their offspring. The irony of all this was that all the while the key insight and its experimental confirmation had been worked out by a quiet monk and published in a scientific journal where anyone might have read it.

In the selection that follows, Ruth Moore tells the story of that monk, Gregor Mendel, and his work. In 1900, 34 years after Mendel's research was published, its importance was finally realized and follow-up work was begun. This was the beginning of the science of genetics.

Mendel's work indicated that inheritable traits were transmitted in discrete units which could only combine in certain ways. Because the units (he called them "factors," we would call them "genes") were discrete entities, they persisted in the population until the last living being containing them died without passing them on. Darwin had thought that inherited traits would blend together in the offspring and not remain discrete. Were that the case, it is difficult to see how the population could ever *evolve* because it would always be tending back to the average traits. But if inherited traits persisted, then those with survival value could eventually become a significant percentage of the "gene pool" of the population.

Mendel never knew exactly what a "factor" or "gene" was, only how it behaved from generation to generation. For that nearly a hundred years had to pass from when Mendel began his work—until 1953 when James Watson and Francis Crick discovered the structure of the DNA molecule. Then it was possible to see that the "factors" passed on were really codes embodied in the protein structure of DNA.

Gregor Johann Mendel was born on July 22, 1822, into a long-settled peasant family living in the tiny village of Heinzendorf in what was then Austrian Silesia and is now Czechoslovakia. As a boy he aided his father in the grafting of trees and in cultivating the garden that was a part of "peasant holding no. fifty-eight." He received additional training in horticulture and beekeeping at the village school. Johann did so well in this work and in his academic studies that his teachers urged that he be sent on to the high school at Troppau. Mendel never lost his love of growing things as he struggled to make his way at Troppau and later at the Philosophical Institute at Olmütz. Gradually this early interest turned into a wider interest in science, although little science was included in the institute's course....

Mendel was so excellent a teacher that the head of the school confidently urged him to try for a teacher's certificate. Mendel wrote the necessary essay and presented himself in Vienna for the oral examinations in the natural history and physics that he wanted to teach. He failed to pass the examination, partly, it seemed, because he lacked certain formal knowledge of his subjects. In part his failure may have been due to the conviction of the friendly chief examiner that the unusual young teacher should be given an opportunity for "higher scientific training" at the University of Vienna.

The abbot of the monastery agreed, and Mendel went to the university for four terms. At the end, he returned to teach in the Brünn "Modern School."

Soon after entering the monastery Mendel began trying to develop new colors in flowers. He also took up the breeding of gray and white mice. Although the latter experiments were dropped after a while, probably at a hint of ecclesiastical disapproval, they were a certain indication that the work with flowers was not a fancy but the first approach of a serious scientist to a fundamental problem.

Mendel's own account of how he happened to begin his studies is a simple one. As he tried to produce new colors in his flowers, he acquired considerable experience in artificial fertilization. His interest was aroused by the surprising and unaccountable results that he sometimes obtained. Whenever he crossed certain species, the same hybrid forms cropped up with striking regularity. But when he crossed one of his hybrids with another, some very different characters sometimes appeared among their progeny.

These phenomena sent Mendel to the literature. He searched all the books available in the monastery and elsewhere in Brünn. A number of scientists were working with the problem of hybrids, but no one had discovered any law governing their formation—if, indeed, anyone could conceive of a law capable of explaining the infinitude of forms in the multitude of plants, Inheritance in all its profusion seemed beyond any nailing down.

Mendel nevertheless had seen certain forms appearing with regularity. Others too had noted this phenomenon. They had not, as far as Mendel could find, counted the forms and classified them. Nor had anyone arranged hybrids according to their generations or worked out their statistical relationships. Mendel, in his remote Austrian town, did not realize that even his concept of counting and figuring the mathematical relationships of hybrid plants was original and entirely his own.

To study heredity properly, Mendel saw, a large number of generations would have to be bred. He also recognized that many plants would have to be included in each generation. A few might produce a misleading picture. "It indeed required some courage to undertake such far-reaching labors," Mendel wrote later in the introduction to one of his monographs. "It appears, however, to be the only way in which we can finally reach the solution of a problem which is of great importance in the evolution of organic forms."

The year was 1854. Many people were asking how the great number of species could have arisen, and yet Charles Darwin would not publish *The Origin of Species* for another five years. It seems unlikely that the Austrian monk was planning a direct study of the great problem of evolution. His interest was concentrated on heredity and how traits could be handed on from parent to offspring. Nevertheless, as his words show, Mendel was well aware of the evolutionary significance of the work he was about to launch.

Mendel planned the big experiment with care and foresight. His first requirement was a plant with varied characters, each of which bred true to form. If a plant produced green peas, he had to be sure that it would go on producing green peas if crossed with another green-pea plant. Mendel also had to have a plant that could be protected from foreign pollen during its flowering period. If even one bee carried in some pollen other than that he would dust on, the whole experiment would be upset. Equally important, the characters had to be easily observable and countable. It was a huge task Mendel was laying out for himself....

...From the thirty-four varieties [of seeds] Mendel selected twenty-two, and from these, seeds with seven sharply contrasting paris of characters. As he wanted to make no fine-line decisions about whether or not a part of a plant had changed, he selected the following clear-cut differences with which to work:

1. The form of the ripe seeds—round or wrinkled;
2. The color of the peas (the seed)—yellow or intense green;
3. The color of the transparent seed cover or skin—white or grayish;
4. The form of the ripe pods—inflated or constructed between the peas;
5. The color of the unripe pods—green or yellow;
6. The position of the flowers—distributed along the stem or bunched at the top; in other words, axial or terminal;

7. The length of the stem—tall (six or seven feet) or dwarf (three-fourths to one and a half feet).

* * *

Each kind of seed went into its own jar—to await the spring of 1856.

Mendel was at last ready to launch his experiment proper. In one section of his long narrow garden he planted round seeds, and in the next section, wrinkled ones. In other sections went the yellow and the green, and, in their own places, all of the others. Keeping the characters that he wanted to compare next to one another would simplify the work that lay ahead...

As soon as the pollen had ripened in the...round peas, Mendel collected a bit on a fine camel's-hair brush and, removing the bags on the wrinkled peas, dusted it on their stigmas. The little bags then were tied on again to keep away the bees and other pollen-carrying insects. To make certain that his experiments were not affected by which plant served as the seed parent, Mendel also reversed his fertilization procedure. Some of the pollen from the wrinkled peas was deposited on the prepared stigmas of the round...

His first indication of what was happening would come when the pods formed in the "unripe pod color" section of his garden. As the pods appeared and filled out and grew longer, Mendel saw with elation that all of them were green. Whether they grew in the "yellow" half of the plot and came from parents that had produced yellow pods or in the "green" half and sprang from parents with green pods, all of them hung green on the vines. Yellow and green parents alike had produced green offspring. Mendel searched his plants. He could not find a single yellow among the unripe pods...

Confirmation of this striking result came as the summer drew to a close and the pods dried on the vines...

During the winter, as he studied his jars of peas and the way all had completely taken on the character of one parent, it was evident that one characteristic in each pair had been entirely "dominant" over the other. Mendel decided to designate the hereditary trait that prevailed as a "dominant." Later he named the other factor, the one lost to sight, a "recessive."

Only the first step had been completed in the big experiment. The next, the crossing of hybrid with hybrid, had to await another spring. As the sun grew warm and the spring rains fell, Mendel planted his hybrid seeds. Each carefully marked group went into its own plot, but this time his procedure would be different. He would not operate on the buds, He would permit the peas to fertilize themselves in the natural way, and thus would obtain a cross of two identical hybrids.

The plantation was again filled with pea vines, and Mendel waited as patiently as he could for the pods and peas to form and dry.

As the young pods grew, Mendel saw a few yellows appearing among the greens. He counted both colors and entered the figures in his notebooks. When the pods dried in the "pod form" plot, some were crinkled around the peas they contained; they were, in the terms he was using, constricted. The others were inflated. Again Mendel counted both types, and again he reserved judgement...

Mendel continued his plantings through six or seven generations in all cases. As he did so, he learned that the 3-to-1 ratio applied to the appearance of the peas only. When Mendel planted the hybrid parents, they produced both round and wrinkled seeds. One fourth were pure rounds and, as long as they were planted, would yield only round peas. Two fourths were round in appearance, but were actually hybrids and in the next generation would produce round and wrinkled peas in a steady 3-to-1 ratio. The wrinkled peas from the union were again pure wrinkled and never would yield anything but their own kind. The wrinkled peas from the first hybrids yielded only wrinkled peas and continued to produce only wrinkled as long as he planted them and permitted them to fertilize themselves. They were pure recessives also...

"It can be seen," said Mendel, "how rash it may be to draw from the external resemblances conclusions as to their internal nature."

Appearance meant little or nothing. It was not surprising that human beings had eternally been bewildered when they tried to understand heredity. The surface differences of offspring were legion in themselves, and yet below the surface lay still other differences and possibilities for variation.

With the clarity and simplicity of genius, Mendel labeled the "dominant" in the union with a capital

Ruth Moore

A, and the recessive with a small *a*. A constant dominant thus would be formed by the coming together of two *A*'s, and it would be described as *AA*; a hybrid by either *Aa* or *aA*; and recessive by *aa*.

With this understanding, it was possible for him to chart this coming together and separation and the clear-cut results that it produced. Mendel drew, and the world has since followed, his chart of heredity...

Charles Darwin also had experimented with peas and had noticed that the hybrid divided 3 to 1, but Darwin was no mathematician and did not pursue this revealing indication of order in heredity. Mendel, on the other hand, was an excellent mathematician. The new experiment confirmed what he had already glimpsed in his first experiments. He was simply obtaining every combination that could be formed by the separate factors present. If A and *a* were combined, only one combination could be formed: *Aa*. If *Aa* and *Aa* came together, four combinations could be made: *AA*, *Aa*, *aA*, *aa*. And this was exactly what had happened...

Mendel had no way to look into the egg and pollen cells to search for the hereditary factors whose existence he inferred from his experiments. His results, however, were explainable in no other way. Just as Mendel had grasped the secrets of the combinations of these factors, so did he reason out the biological basis that had to underlie them. He formulated three laws of inheritance:

1. All living things are a complex of a large number of independent heritable units.
2. When each parent contributes the same kind of factor and the two come together in the offspring, a constant character is produced. But if one parent contributes one kind of factor, say *A*, and the other another, say *a*, a hybrid results. When the hybrid forms reproductive cells, the two differentiating elements "liberate themselves" again and thus are free to form new combinations in the next union.
3. The factors are unaffected by their long association in the individual. They emerge from the union as distinct as when they entered it.

In a letter to a fellow scientist Mendel enlarged upon this radical idea: "The course of development consists simply in this: that in each successive generation the two primal characters issue distinct and unadulterated out of the hybridized pair, there being nothing whatever to show that either of them had inherited or taken over anything from the others."

This was Mendel's ultimately famous law of the purity of the gametes, or reproductive cells. Mendel regarded it only as a hypothesis, and felt that he must subject it to additional tests...

At last the time had come to report on his eight years of unremitting work. during the fall and early winter of 1864, Mendel checked and rechecked his findings. In his fine copperplate script he wrote the paper that eventually would explain to the world the all-important phenomenon of heredity.

The Brünn Society for the Study of Natural Science was holding its February 1865 meeting at the Modern School, the school in which Mendel taught. The night was cold and snowy, but most of the forty members came to hear Mendel report on the work he had so long been carrying on at the monastery. Curiosity about it was keen.

As Mendel read, this curiosity gave way to incomprehension. The several botanists in the society were as much confused by the report on invariable hereditary ratios in peas as were the other members—a chemist, an astronomer, a geologist, and an authority on cryptograms. Mendel spoke for the hour allotted to him, and then announced that at the next meeting he would explain why the peculiar and regular segregation of characteristics occurred.

The next month, as Mendel presented his algebraic equations, attention wavered. The combination of mathematics and botany which Mendel was expounding was unheard of. And the idea that lay behind it, that heredity was a giant shuffling and reshuffling of separate and invisible hereditary factors, stood in such diametrical contrast to all that had been taught that it probably could not be grasped. No one rose to ask a question. The minutes recorded no discussion.

A number of the members spoke to Mendel afterward about his experimental work. "I encountered various views," Mendel told his friends. "No one undertook a repetition of the experiment."

The editor of the *Proceedings* of the society extended the usual invitation to Mendel to prepare his paper for publication in the society's journal. "I agreed to do so," said Mendel, "after I had once

Ruth Moore

more looked through my notes relating to the various years of the experiment without being able to discover any sort of mistake."

Mendel's monograph—*Versuche über Pflanzenhybriden*—appeared in 1866. According to the society's custom, copies of the *Proceedings* in which it was included were sent to Vienna, Berlin, Rome, St. Petersburg, and Uppsala. But the brief paper that could have altered all ideas of heredity, and at a time when Darwin was still at work on the role of heredity in evolution, attracted no attention. It sat all but unread on library shelves.

From Ruth Moore, *The Coil of Life (New York: Alfred A. Knopf, 1961).*

Questions and Topics for Discussion and Writing

1. Why was it so important that Mendel thought to count and record the characteristics of each generation of plants?
2. Explain the concept of dominant and recessive genes. Explain how the 3-to-l ratio in visible characteristics comes about.
3. Why, in your opinion, was no notice taken of Mendel's work for 34 years, despite its great importance for evolutionary theory (which *was* being given lots of notice)?